I am deeply grateful to those who have freely given
those for whom, the recounting of memories of certain
by the recounting of such times.

Everyone who has contributed is acknowledged,
mischance your name has been omitted, please forgive

Every effort has been made to ensure accurate rec(

◆

For my lovely, and loving, Grandchildren,
Helen, Emma, Samantha and Victoria.

◆

Copyright © Harold Cox 1994.
ISBN No. 0 9523332 0 1

Printed by Parker & Son (Printers) Ltd., 70 Guild Street, Burton upon Trent, Staffordshire DE14 1NE
Telephone: 01283 568820   Facsimile: 01283 510566
Email: office@parkerandson.co.uk

CLASS C DESTROYERS

GREYHOUNDS OF THE SEA

INTRODUCTION

The word destroyer, used to describe a ship is a fairly modern one, derived from the need to combat the torpedo boat, a very small, very fast (and also sneaky and ungentlemanly according to its many detractors), type of craft which could sink the largest and most powerful ship then in service.

Torpedo boat destroyers were born. Shortened to 'destroyer' the term is still in Royal Navy use today.

Initially of several hundred tons, a modern (1994) destroyer is of some 5000 tons, and bears little resemblance to the pre-war, or war-time vessels. Usually carrying 4 guns of 4.7 inch calibre, together with torpedo, depth charge, and secondary armament, the 1930's Royal Navy destroyer, was aptly described as a 'Maid of all work'. Speedy ships, they could handle almost any situation at sea, being handy, if uncomfortable seaboats.

In September 1939, Britain and France went to war against Germany who had invaded Poland, after occupying several other states. After a fairly quiet start to the land war, the Germans attacked through Belgium and succeeded in overwhelming the Allies, the British evacuating from Dunkirk and other areas, leaving masses of equipment behind. Norway and Denmark followed, thus giving the Germans use of the whole European and Scandinavian seaboard, some 2200 miles long. Britain, now alone, could not rely on Southern Ireland as a sea or airbase, and thus faced a most desperate situation.

Prime Minister, Neville Chamberlain, in 1939, thought he had secured a peace with Hitler, this proved a false hope. Succeeded in 1940 by Winston Leonard Spencer Churchill who presided over a Government formed of a coalition of parties, and who was to tell the House of Commons, and therefore the country:- "Death and sorrow will be the constant companions of our journey, hardship our garment, constancy and valour our only shield, we must be united, we must be undaunted". On the 8th October 1940, when he uttered those words, the British people knew he was right, and that he was not offering any easy solutions.

Against this background, and amid grim news of U-boat successes, one of the decisions taken was to construct a series of new destroyers, the first of which was to be the 11th Emergency Destroyer Building Programme, consisting of 8 ships, the technical details of which are to be found later on in the book. Two ships were laid down in 1942, the remainder in the next year when ships of the 12th. and 13th. E.D.B.P. were also commenced. Four more were commenced in 1944. Twelve vessels had their original name changed, this served to regularise the naming since the first eight names all now commenced with the letters

CA, the next eight were CH, the third flotilla CO, and the fourth CR. One further flotilla was envisaged commencing with CE, but whilst CELT and CENTAUR were built, they do not form part of this book. CAESAR was to be the name ship of the CA's, but CAPRICE was the first to be completed, and the only one in commission by D-Day 6th. June 1944.

These ships represent the end of an era. Royal Navy men of that period will say without any hesitation that they "Looked right, they looked the part". Never mind that they were overcrowded, very uncomfortable in bad weather, open bridges, no protection for guns crews other than the front shield. Nominally crewed by 186 men, this number rose in wartime, and when carrying Flag Officers and the rest. Of necessity, officers and ratings lived in much closer proximity to each other than was the case on large ships. Normally Royal Marines would not be carried, and Bugles not used. The quartermaster would move through the ship piping orders by using the very old method, the bosuns call.

These are the ships, with their most important addition, the men, who sailed and fought them. We will meet each of them in turn, but their position in the book reflects neither superior, nor inferior work or service on the part of any.

◆

CAPRICE refuels in China Seas.

CONTEST

Laid down on the 1st, November 1943, CONTEST was launched into the Solent from J. S. White's yard in the Isle of Wight on 16th. December 1944 whilst the Allies were fighting a sudden, unexpected attack by the Germans in the Ardennes, and known from thereon as the "Battle of the bulge". Completed nearly a year later on 9th. November 1945 and therefore taking no part in the war that had lasted for nearly six years.

She was not to be wasted however, in early 1946 she was in Far Eastern waters, visiting Shanghai with COSSACK, Japanese ports, and Nanking where Britain retained an embassy staff. CONTEST carried the British Ambassador Sir Ralph Stevenson at one stage. Late 1947 and she was homeward bound. During the trip home from Hong Kong she took part in a simulated attack on Singapore, bringing back memories of the Japanese attack from landward in 1942, when General Percival surrendered his army. Ships in company included FINISTERRE, and the Light Fleet Carrier GLORY. Further work as attendant destroyer to the Home Fleet aircraft carriers followed, and in 1950/1951 she acted as target ship for submarine training purposes, usually based on vessels from H.M.S. DOLPHIN the shore base at Gosport.

Based on the Scottish Western area she carried out similar duties and would track submarines and throw over a small charge at the appropriate moment to signify they were on target, the submarine commander naturally would always hope to hear the explosion as a very distant sound effect. The speed of sound in water is more than 4 times than in air, a case of news travelling fast!

Frank "Slinger" Woods had joined the Navy in 1942. He describes himself at that time as: "17, Green as grass, 10 stone wet through". Drafted to a 1000 ton trawler as a stoker, he found that she was a coal burning ship, with "A 4" Popgun at the bows and twin .5" machine guns aft." After several journeys, including one to Reykjavik in Iceland, he was drafted ashore to Portsmouth. Late in 1943, he and a great many others including R.N., R.M., and R.A.F. personnel found themselves at a huge base in Scotland there to be kitted out in khaki and told: "You're in the army now!" He eventually served in Europe and Italy in the Royal Engineers. He found a great difference in life as a soldier, compared to serving on the trawler MARIA, ("She stank of fish - we stank of fish") he remembers.

Finally released from the army, whilst on his last leave he again joined the Navy. Fated, he thought to remain with the 'Black stuff', his first R.N. ship was a survey vessel of the coal burning variety, H.M.S. CHALLENGER. Next came a Tank Landing Ship, NARVIK, Slinger had a trip to the West Indies in her, on return he volunteered for submarine service, accepted, he did 8 weeks training at DOLPHIN before being returned to General Service due to a recurring medical problem.

On the 3rd. January 1950 he left Portsmouth Barracks with a draft chit in his hand for H.M.S. CONTEST berthed at North Wall Jetty.

Frank takes up the story: "I jumped down from the wagon and saw CONTEST looking a total mess, having just undocked after a refit. Up the brow, at the top stood a matelot in the biggest greatcoat I had ever seen, collar right up above the top of his ears. 'Wot is you' he said, fingering the bosuns call hanging on a long silver chain around his neck.

On being told that I was joining from RNB he directed me toward where I could find the Chief Stoker, never having been on a destroyer before this was not as easy as he had indicated. After much scrambling round the ship I finally met up with the man in question, he looked at my papers and said 'you've not done much have you'? This reaction was to be evident a great deal in the next few days, brought about no doubt by my 4 campaign medals which would give the impression of a great deal of R.N. time. Go down to 7 Mess and find Leading Stoker Tosh, he'll fix you up.

The Chief had pointed vaguely forrard, so I made my way down and toward the sharp end, kitbag on one shoulder, hammock under the other arm, 4 times I'd done this, should be used to it now. After one ladder realised that I was in a seamans mess, down the next iron ladder, 7 Mess? yes mate, this one here!

One single fouled anchor on his arm told me he was a killick, was he Tosh? Writing with a piece of pencil about 1½" long, labouriously totting up a few figures, muttering to himself, I asked if he was Tosh. 'What if I flickin' am' he growled without looking up, the Chief Stoker told me to report to you on 7 Mess, I said. The man was huge, he swivelled his head around, his face was carved out of stone, deeply furrowed brow, and he looked at me from pale watery eyes that had seen a million miles of ocean, 'spect your Janners relief, right, this is your locker'. Another vague gesture toward the ships side, the third since I got aboard, followed by, 'and them's your mick 'ooks' this time pointing at the 2 small indented areas of the metal bars which had supported many hammocks, and which were to be mine. His head dropped and his slow writing commenced once more, after a few seconds, the great head raised itself once more 'Wait a mo' he fixed me with his eyes like a dog fixes a rabbit 'Are you G. T. or U.A.,?' (Grog, Temperance or Underage). On telling him that I was Grog an immediate if slight, melting took place, a little of the aggression had gone away. 'Well nar' he went on, you'll be Port Watch, stow your gear, and put your mick in the nettings'.

My hammock stacked away with all the others, I unloaded my kitbag lifted the locker lid, which like all the others provided the seating alongside the ships side, and began to stow my gear. Sensing Tosh was taking an interest I looked up just as he said 'Oh for Christ's sake gerroutof the way, this is your first small ship'? I nodded, 'thought so, we get some right useless buggers nowadays, not like it was,' and as if to emphasise, 'not like it was'. Very quickly he had restarted to get my gear away, a few little grunts, and a final 'Gawd help us if we ever have to go into action' and most of my stuff was packed. That locker was no more than 2 feet by 2 feet, 6 inches.

Having joined the mess, I now had to join the ship. Tosh, still sighing gently 'Go up top, find the Cox'n, tell him you're Port Watch, awright' having said that his head went down once more, pencil poised for his next literary effort, he gave me every indication that Slinger Woods, Stoker 1st. Class, Royal Navy, did not exist.

The Coxn's office was in the canteen flat, I knocked, 'Come in and be flickin damned' what a welcome thought I. After telling him who I was, he asked me a few questions, I informed him that I was Port Watch, 'Noted, here's your Station card' he handed me the little card which effectively identified you as a member of CONTEST'S crew, stated your watch, whether G,T, or UA, action station, and rating. 'Oh, you're G' he said, again with that ever so slight warming of the voice I'd noticed whilst with Tosh. UA, underage meant you were not entitled to rum, you could not help this. Temperance however was a totally different kettle of fish, if you were the only T. rating on a G. mess, you were beyond the pale, the runt of the litter, the black sheep, and you would be treated accordingly.

Business concluded I turned to go 'Wait a minute, You've been around' the Cox. said looking at my ribbons, a bit Chief, I replied, 'Not too many newcomers have nowadays' saying this he turned away and I knew that this time he really had finished. Closing the door I thought I'd set off an earthquake, it turned out to be dockyard men hammering or rivetting up top, after one hour aboard, it seemed like a year. Picking my way over cables, and other assorted pieces of dockers equipment I went down to my new home.

'UP spirits' was piped, 'Hands of the mess for rum', bit late thought I, Tosh was already half way down the mess ladder with the neat, round stainless steel mess tin with its quota of grog for 7 mess, probably first in the queue, an aura of rich, molasses wafted across toward me, before I could speak he jerked his free hand toward the small stowage box on the bulkhead, 'Right you, get the rum tackle ready on the table, don't forget the flickin oilcloth either'. Oilcloth duly on the table, Tosh measured each tot out, bodies appeared in a bunch from the deck above, the quiet mess was now transformed, a noisy group took their tots. Two men carrying food appeared, both in number 8's with stoker's badges, on the arm, they proceeded to dish out the food. Seeeecure, hands to dinner, men under punishment to muster' was the next pipe, needing no prompting, ravenous, I scoffed my dinner, if it had been dead dog I would have eaten it. The 2 blokes who'd carried the food down now began to wash up and clean the table, cooks of the mess for the day, they would be very busy with that additional work on top of their normal duties. One of the 'cooks' made away with the large stainless steel mess tin with the waste, at sea that would go over the stern, but in harbour it went down the brow for pigswill.

'Out Pipes, Hands carry on with your work; Leave. Leave to the 1st. part of the Port watch, and 2nd. part of Starboard Watch from 1600 to 0745. Chief and Petty Officers 0815. - Both Watches of the hands, fall in'. Dinner was over. Reporting once more to the Chief Stoker he took me down to number 2 Boiler Room, there to be introduced to the Petty Officer who, for the forseeable future would be my judge and jury. Within one minute he had informed me, and all others within earshot, that 'they', I assumed he meant My Lords Commissioners of the Admiralty, sent him 'flickin useless prats'. 'What was your last ship' he asked me belligerently, the NARVIK I explained, 'what the flickin hell was that?', a LST I said, feeling rather guilty about the NARVIK, but not sure why. 'Jesus Christ and all his flickin disciples, don't you know anything about a proper ship? Get stuck in with the boiler clean' At that he turned towards the little throng who had stopped to hear what was said, waving his arm towards the manhole door in the furnace side 'Now then you bastards get crackin'. Following the others into the furnace via the 18" square manhole cover I realised immediately that I was never going to enjoy boiler cleaning, whilst not claustrophobic, I didn't relish being cooped up in a dimly lit, smelly dusty furnace. Re-assuring myself that I would certainly not be the first - or last - individual to feel that way, I joined with the lads and set to work.

All servicemen grumble, but the rejoinder is always the same from your messmates 'Serves you right, you shouldn't have joined if you can't take a joke'. Chief Stoker, asking me the next day how I was getting on remarked 'Don't worry about Frank (the PO who had harangued me a bit in the Boiler Room), that's his way but he's got a heart of gold really'. Later on I was to realise this was so, and he also knew his work from A to Z.

After a period of service, I was taken off the Watchbill and given a daywork job, known to the rest of the mess as 'an upper-deck stoker', I carried out a number of tasks keeping an eye on oil and water levels, looking after servo-mechanisms, and manning the upper deck capstan after requesting steam to supply the capstan from number 1 Boiler Room. Shortly before going to sea in a warship, 'Special sea dutymen to your stations', is piped and my work at the capstan to raise the anchor or take in the cable would commence. Left alone, once competent and reliable, the work suited me, and I took on the extra work of playing records on the SRE, and even used to go ashore buying records for the ship. Anytime that leave was piped, I could go ashore! The day arrived when I faced the Board for promotion and whilst with CONTEST became Leading Stoker.

Discharged from the Royal Navy on 16th. January 1952, on medical grounds, I left the service for the second time."

Dennis Barker of Leeds served aboard CONTEST as an ASDIC operator. Whilst at sea, they received a signal to sweep in a particular area for a submarine which was reported overdue. H.M.S. AFFRAY was carrying out exercises with dockyard staff aboard, when she failed to make a radio report, 'Submiss' procedure was instituted. It should be remembered that the signal 'Submiss' may be made due to a failure of radio communications the vessel being in no danger, those ashore cannot be certain of events purely because a radio report had not been received.

Sadly, in this case which occurred in the English Channel, as time passed it became more and more likely that something untoward had happened and CONTEST, later joined by others including a lifting vessel capable of moving very heavy weights, became part of a sizeable taskforce searching for AFFRAY.

Dennis, the only ASDIC operator aboard out of the 4 normally carried, (One, the ship's postman was ashore collecting mails, one was on compassionate leave, and one was sick) was therefore hard pressed to keep going with his watchkeeping duties. Leaving the ASDIC cabinet only to answer the calls of nature, his food and drink delivered to him, he attempted to train another seaman to carry out sweeps with the submarine detector gear. Tiredness eventually engulfed him, he remained in the ASDIC cabinet however with the headphones round his neck and reckons that even when asleep he would be awakened by the Ping from his set, returning as a Ping - er. This situation lasted for two days says Dennis.

H.M.S. AFFRAY had indeed gone down taking all hands with her. After she was found, TV cameras were able to confirm that it was she, her name showing clearly on the surface television screens. The so called Snorkel breathing tube, which enabled the submarine to run submerged and still run the diesel engines, appears to have collapsed and the boat would have flooded up. The price of Admiralty ....

Dennis Barker remembers whilst based at Plymouth, CONTEST was detailed to carry out test firing of shells fitted to a new and secret fuse. Incidents arose of shells exploding prematurely due to too sensitive fuses activating the explosive charge when in rain, snow or spray. Whilst this seems quite extraordinary in the 1950's, similar effects have been noted as recently as 1993 by British forces using Italian made fuses. CONTEST duly sailed from time to time when weather conditions or sea state, or both were suitable and carried out the secret tests. Shore leave would be given as appropriate, and Dennis recalls the astonishment of himself and oppo's when in Plymouth someone looked at his cap tally 'CONTEST' and said 'Oh you're the lads testing that new fuse then'? So much for secrecy!

On 15th. June 1953, the Queen and Prince Philip reviewed the British, Commonwealth and Foreign ships anchored in the Solent. CONTEST took part in that event. Refitted with minelaying equipment fitted on each side of the quarterdeck, she joined the 6th. Destroyer Squadron on Home and Mediterranean duties. Placed in reserve in 1959, she made her way in early February 1960 to Grays, where she was broken up.

◆

*One Division of the 8th. Destroyer Squadron.
Nearest camera: D20, COMUS, then DO3 CONCORD, D34, COCKADE, D90, CHEVIOT (Captain (D).*

Heard in the petty Officers mess:-
The Chief Stoker says he wants waking on pay day whether he's had enough sleep or not!

COMET

Built by Yarrow of Scotstoun, COMET was launched on the 22nd. of June 1944, just 16 days after the Allies had landed in Normandy. Cherbourg was being attacked, and the siege of Imphal in Burma was lifted. March 1947 and her duties took her to Tsingtao in the North of China, a seaport in the Yellow Sea betwen Korea and China. Leaving for Japan in company with Captain (D) in COSSACK she was detached and ordered to assist the KOHUK MARU at Barlow Island. Amongst many ports, she visited Sasebo in Japan in company with COSSACK and COMUS.

Returning to the United Kingdom, she was prepared for further service in warm climes, and sailed to Malta for working up exercises in 1949. Undergoing the vast array of tests, trials and exercises required to bring her and her company to full efficiency, she was at sea on night encounter exercises with other Malta based destroyers when, in darkness and without lights, an order was misunderstood and COMET was struck on her Starboard side by CHEVRON. COMET heeled well over to Port, shuddered, lost way, and eventually regained a reasonable trim, not least due to the efforts of all available hands using pumps and bucket chains to remove the invading sea water. When a tug arrived, measures were put in hand to tow her back to Malta where temporary repairs were carried out, prior to her return to Britain - necessary to deal with the damage sustained in the collision.

COCKADE was made ready and left the U.K. to take up COMET'S place in the Far East.

During her 1954 - 1955 commission COMET was once again in the Mediterranean, her Commanding Officer now was Commander Burton, known to the lower deck at least as Black Bob, with some affection it must be said.

Gamal Abdel Nasser, an Egyptian army officer, and Major-General Mohammed Neguib were jostling each other for the ultimate control of the Egyptian State. In October 1954, Britain and Egypt signed an agreement, the main thrust of which was that Britain's forces would leave the Suez Canal Zone by June 1956. Ever watchful, and wary of Nasser's impatience, British warships began to escort convoys through the canal to forestall any difficulties. After the eventual withdrawal by Britain, Nasser announced that Egypt had nationalised the Suez Canal, thus provoking the Suez War. COMET took part in these escort duties.

Sergeant Mike Thomas serving with the army in Egypt at this time recalls several R.N. ships visiting Port Said. Walking along the waterfront one day he noticed one of the young lads who frequent the area, often scrounging for "tickler's jack" - tobacco, or food, and who he recalls being tutored by a sailor to speak, recite and even sing in English. Seeing two smartly dressed officers approaching, and clearly wanting to impress, he broke into song. Whether he realised that one of the officers was the navigating officer, the other, and more to the point, the R.N. Chaplain, is not known. The little Arab struck up to a well known tune Mike remembers:

> Jesus wants me for a sunbeam,
> and a bloody fine sunbeam am I ....

Mike Thomas reckons no words could describe the change which came over the padre's face during those brief moments.

Bill Johnston, of Newbridge recalls the canal escort duties, COMET leading a convoy on its steady progress toward the Great Bitter lakes.

Never short of work, the British forces now had another task to perform when the Greek Cypriot people wished to join with their Greek neighbours on the mainland. Enosis - union with Greece - was to be fought for, generally by Eoka a terrorist organisation, which at one stage exerted far more power than its small numbers would perhaps have warranted. The Troodos Mountains offered lawbreakers a wonderful hiding place, and many areas of Cyprus were wild rock-strewn tracts. Many helicopters would conceivably have transformed the situation for the British commander, but he had to work with what he then had.

H.M.S. COMET helped the ground forces by patrolling the coasts to keep undesirables out, and to attempt to control gun-running. Some evidence of the success of this is a picture that Bill has showing 3 gun-runners aboard COMET after being arrested offshore. However Cyprus has a very long coastline and effective blockading would be most expensive in the use of ships and aircraft. As in the Palestine Patrol situation 10 years earlier, some inevitably would get through.

COMET took on a new role, losing some armament, and gaining a consignment of sea mines together with the necessary 2 rails at the stern which enabled the mines to be laid.

Suez followed, many ships had orders for bombardment duties in support of the landings at Port Said, but it is thought that the very successful operation by the air and land forces removed much of this requirement. World, and particularly United States opinion, determined that the Israeli, French and British forces withdrew from the canal zone, and Nasser once again had a smile on his face.

COMET was broken up at Troon commencing 23rd. October 1962.

◆

Suda Bay, Crete. Guard of Honour for Governor.

That package of assorted miseries which we call a ship ....

CARRON

The Luftwaffe was bombing Malta relentlessly, in a period of 27 days, Britain's fortunes at sea received a massive blow when the Eisbär (Polar Bear) wolf pack sank 28 ships off the East and Southern coasts of Africa. During this period on the 11th. September 1942, Job number J1141 was laid down at Scotts of Grennock. Originally to be named STRENUOUS, she was launched on 28th. March 1944 as H.M.S. CARRON.

Completion date arrived, quite quickly in her case, in early November 1944, and she joined her sister ships in the Home Fleet 6th. Destroyer Flotilla. Service in the North Atlantic and Norwegian waters followed until May 1945, now wanted for service against the Japanese Empire, she slid into dock for a short "warm climes" refit.

Now a familiar routeing for R.N. ships at the time, she followed many others to Columbo, there to start preparations for the Far Eastern conflict. Two atomic bombs changed the whole scene and she found herself, and nearly everyone else at peace. One snag however was the unrest in the Dutch East Indies, the British services were still employed in that area assisting the Dutch in regaining control from rebel forces, CARRON helped by evacuating refugees. Bombardment of rebel positions was required and CARRON was able to offer her support in this way.

The band played change partners once again in British affairs and the wholesale movement of ships and men to Britain began. CARRON returned home and went into Category 'B' reserve at Chatham.

During the years 1953 to 1955 the ship underwent a very sizeable reconstruction. One set of torpedo tubes was removed, as was one after 4.5" gun. Her bridge was improved, 2 Squid mortars fitted, 2.40mm Bofors, and new power operated gunfire control systems were also installed. At this time her wartime complement would hardly have recognised their old ship!

By 1958 the ship had neither 'B' or 'X' guns from the original four, she joined the Dartmouth Training Squadron, after which all her 4.5" guns were removed and she joined the Portsmouth Training Squadron in July 1960, attached to the navigational establishment H.M.S. DRYAD.

In 1963 CARRON was placed in reserve, and she finally was taken out of service and scrapped by Ward's at Inverkeithing commencing on 4th. April 1967. The 'C' Class was now down to the last few survivors. Strange to relate, apart from 1 CH - CHEVRON - they are all CA's, the first to be built, and of rivetted construction, whether that is in any way significant, only the ship-building experts will know!

CARRON has the battle honour "Arctic 1944".

House of Commons. Winston Churchill, Prime Minister at that time meets Jennie Lee, another Member of the House: Winston, you're drunk! Churchill: I know, but you're ugly, and I'll be sober in the morning!

CAVENDISH

John Brown's shipyard saw the launch of CAVENDISH, ex SIBYL on April the 12th., 1944, in a week when 2 U-boats were sunk, and the Russians recaptured Odessa. Listed as Job number J1606, she had her keel laid as a leader, and was the 7th. CA to be completed.

CAVENDISH in company with the escort carriers NAIRANA and CAMPANIA sailed from Scapa Flow to attack enemy ships off the Norwegian coast. They were joined by the cruiser BERWICK and the Canadian Destroyer ALGONQUIN. Four smaller ships were attached to this formidable force. On arriving at the scene, in good visibility, the attack was called off after what seemed to be a short time by some members of CAVENDISH'S ship's company, although some success was reported. Scapa Flow was reached on the last day of January.

They were soon at sea again, this time they rendezvous with Rear Admiral R. R. McGrigor's force escorting a convoy from Greenock to Russia. The merchantmen went to Murmansk and the warships to Vaenga where a Lieutenant Royal Navy was the Commanding Officer. Sailing for the return trip home, one of the escorts was damaged by a torpedo, and a steamship in the convoy was sunk by enemy aircraft. The weather was foul for much of the trip and this, it is thought may have helped reduce casualties.

On 10th. May 1945, a U-boat, believed to be U516 surrendered to CAVENDISH.

To prepare to go East now that Hitler had been defeated, CAVENDISH had a short refit at Plymouth. She was in Columbo by early August 1945, just in time to see the Japanese surrender. In the next 9 months she was to visit a great many exotic ports throughout the East Indies and the Pacific Ocean, until on the 20th. June 1946 she was back home in the U.K.

From 1956 to 1959 two Home/Mediterranean tours were carried out, included in these were a visit to London in company with CARYSFORT and CONTEST, and fishery protection duties in Icelandic waters.

Allan Gunnis of Blackpool joined CAVENDISH in Gibraltar whilst she was refitting, he remembers they first turned east on sailing and spent some time in the Mediterranean before returning to the U.K. One further foreign visit and he left her for pastures new in 1964. She became at one time Captain (D) of the 21st. Escort Squadron in Far Eastern waters he recalls. During this time she was also involved in landing men on the Island of Rockall, thus establishing Britain's possession of the outpost beyond all doubt.

Portsmouth was her home in 1964, when she was laid up there. She visited Chatham, and returned again to Portsmouth in 1966.

In August 1967 she was broken up at Blyth. Now 5 ships remained of the original 32 built.

This is the grey funnel line, not your fathers yacht!

COMUS

H.M.S. COMUS was laid down at Messrs. Thorneycroft's yard, Woolston, Southampton on 21st. August 1943. At this time Sicily was in Allied hands. Her hull entered the water on 14th. March 1945, and she completed 20th. December 1946, some 16 months after the end of hostilities, and over 3 years from her initial laying down.

Sailed by the Admiralty for her Far Eastern destination, she joined the 8th. D.F. and was on station by April 1947. She served there until April 1957.

At 0800 on 14th. July 1947 she had an emergency when a man went overboard, fortunately he survived the ordeal. Amongst her early visits was the port of Inchon in Korea which she was to know later when it became part of the hostile enemy coast. Late July and Sasebo in Japan was visited. Late November found her at Number 8 buoy in Hong Kong. Home to the 8th. D.F.

Peacetime exercises and port visits, combined with the usual docking and maintenance work throughout 1948 kept the crew of COMUS busy. One light moment was provided by the regatta at Sasebo when she competed with CONCORD, CONSTANCE and COSSACK. On 9th. February 1950, in company with JAMAICA, KENYA, TRIUMPH, CONCORD and CONSTANCE exercises were held in a wide variety of situations likely to be met with in wartime conditions.

During the Korean War, COMUS, in common with other United Nations ships spent time patrolling the seas against Communist incursion into Allied waters, bombardments as ordered against shore positions were also carried out. Enemy naval forces did not play a large part in the conflict and as a consequence ship versus ship actions were minimal in effect. COMUS assisted at Hill 60 by firing 150 rounds into Korean positions.

Singapore dockyard and a long refit followed her exertions, volunteers travelled to Penang, some 400 miles distant for a period of "leave". In view of the Malayan situation at that time, they each carried small arms and ammunition.

COMUS recommissioned on 14th. July, 1955. During this commission she was signalled to go to the aid of the steamship HELIKON who had reported on 500Kc/s, the main shipping and distress wireless frequency, that she was being attacked by a warship, off Foochow, a seaport on the Chinese mainland some 400 miles South of Shanghai. On reaching her position, COMUS stationed herself between the merchantman and the Chinese Nationalist warship, (of Japanese extraction) and carrying two 4.7" guns, plus Bofors. Numbered 482 on her hull the Chinese signalled COMUS "Get away from this area", at this COMUS cleared for action. It was teatime, 482 fired two rounds of 4.7". COMUS trained her main armament on Number 482 and signalled:- "If you fire again I shall open up". Some time elapsed but eventually the Chinese warship withdrew and the HELIKON went about her business.

In May 1956 during a long refit at Singapore, 10 members of the ship's company sailed a motor fishing vessel along the coast of Malaya whilst COMUS eventually followed them and welcomed them back aboard. Hammocks had been used as awnings to offer protection from the sun whilst aboard the MFV, and the participants found the experience invaluable.

Paying off in November 1956, she was sailed home by the ship's company of H.M.S. MODESTE, arriving in the U.K. in April 1957.

She went to Cashmore's of Newport for breaking up on 12th. December 1958.

◆

D20 COMUS - laying smokescreen.

Damage to CONCORD'S quarter, caused by Korean warship ramming her. Kure, 1952.

Sign over a portal at RNB Chatham: Fear God, honour the King.

CHIEFTAIN

CHIEFTAIN took to the sea in February 1945 as the tide of war was running strongly against the Germans and Japanese. laid down in June 1943, she was finally made ready in March 1946. Built at Greenock by Messrs. Scotts she was ordered to the Mediterranean for service with the other CH's in the 14th. Destroyer Flotilla, soon to become the 1st. She and CHIVALROUS were chummy ships and carried out a lot of work together, quite important in the Palestine anti immigration sweeps carried out.

Despite her name, she was only a ½ Leader, taking over the reins from CHEQUERS when occasion demanded. In 1953 she was placed in reserve, and returned to the Mediterranean in 1955. Suez in 1956 found her amongst about 99 other warships and supporting craft at the Suez crisis. Expecting to bombard Port Said, it is thought that after the air strikes and subsequent landings by the army and marines this would not be necessary in view of the success achieved. Particular attention was paid to the prospect of finding enemy mines, but CHIEFTAIN found none.

In 1957 CHIEFTAIN formed part of the escort to Her Majesty the Queen on her visit to Portugal.

Messrs. Youngs of Sunderland carried out the final act of her lifetime when they broke her up in March 1961.

◆

Haganah ship 'Jewish State' she is believed to have sailed from Bari. CHAPLET shadowing her for a long time before boarding. Unloading at Haifa.

King Farouk, King Farouk, caught his goolies on a hook ......

CHILDERS

Four 'C' Class ships were laid down in November 1943, one of them, CHILDERS, was launched in February 1945. During that week the S.S. HENRY BACON was attacked and sunk by Junkers 88 aircraft, the last merchant victim of the war to be sunk by German air power. In the same week Cologne was besieged, Germany clearly was at her last gasp.

Six countries chose to declare war on Germany during that same week, four of them Saudi Arabia, Lebanon, Syria and Egypt, were in a vital strategic area throughout the 5½ years of the war, in the case of Egypt, the Germans had actually invaded her from the West in the early years! Comment is superfluous!

CHILDERS was constructed at the Dumbarton yard of William Denny, and she was to be the eighteenth 'C' Class ship to be completed.

After completion, which occurred on the 19th. December 1945, she prepared to join her sisters in the 14th. D.F., based in sunny Malta. Carrying out all the duties of a 'peacetime' R.N. destroyer, she was involved in patrolling the Eastern Mediterranean to detect and arrest illegal immigrants into Palestine. Much exercising was the order of the day when these uncomfortable duties allowed.

Manned by Devonport Division men until centralised drafting was introduced, she went into reserve in 1958.

After being laid up at Gibraltar in 1958, she eventually underwent a lengthy tow to La Spezia, Italy for dismantling, the only one of her class so to do directly from R.N. service.

◆

On passage between Marseille and Gibraltar, August 1947. PHOEBE passing stores to Illegal immigrants and 6th. Airborne guards aboard Empire Rival.

Adversity: The state in which a man most easily becomes acquainted with himself, being especially free from admirers then. *Samuel Johnson.*

CONSTANCE

In March 1943, 21 I.R.A. men escaped from Londonderry jail, Northern Ireland (Nothing changes), during the same week a keel was laid down at Vickers-Armstrong yard at Newcastle on Tyne, and CONSTANCE which had been ordered on 24th. July 1942 was commenced. During the same week the submarine THUNDERBOLT was sunk by an Italian corvette in the Mediterranean. Previously known as THETIS, she had sunk once before in Liverpool Bay whilst undergoing trials after construction, and both R.N. and civilian ship-builders men were lost on that sad occasion. Unusually, and almost certainly due to the war situation, she was raised, made ready for sea and commissioned, and put into service.

Adopted by Beckenham in April 1943, CONSTANCE was some two and a half years in building and completion. Launching date was 22nd. August 1944. She was to have a short life of only 10 years.

Commissioned in December 1945, she sailed on 2nd. February 1946 for the Far East and the 8th. D/F. Peacetime duties took her to a great many ports including Shanghai, Hong Kong and Singapore of course, both bases for warship work, many Japanese ports, inlets and bays, Penang, Manila, Ilo-Ilo, Nanking and numerous others.

Between 1947 and 1953 she had 7 Commanding officers: Messrs. Gregory, Gregory-Smith, Baker, Seale, Lyle, Bayly and Morgan. Jack Mead of Newark on Trent served aboard CONSTANCE from November 1947 to 1950. He recalls the Southern Cruise when many of the aforementioned harbours were visited and the ship steamed some 4900 nautical miles. Diverted to Manila in the Philipines, CONSTANCE provided a guard of honour at the funeral of President Hoxa, returning to Hong Kong afterwards.

During December the ex-Japanese destroyer SUMERI was taken to sea, and CONSTANCE in company with others opened fire for target practice, sinking the ship in the process.

Late July 1948 arrived and the ship joined the Northern Cruise visiting Japan and the sea areas to the Northeast of Hong Kong. Much exercising was carried out individually and with other British and Allied warships. CONSTANCE returned to Hong Kong on 24th. September 1948.

CONSTANCE visited Shanghai, prior to entering the Yangtze estuary where she anchored whilst choosing the moment to make the journey up-river to Nanking. Troops in large numbers were to be seen everywhere since the front line between the Red Chinese attacking from the North and the Nationalist Government forces was getting nearer to the area. At 0615 on the 7th. of December, she sailed the 95 miles upstream to Nanking where she would help protect British and other neutrals interests, at that time a standing duty for the Royal Navy. CONSTANCE arrived there at 1515 and commenced taking over from the Frigate H.M.S. AMETHYST. At 0630 the next day, formalities complete, AMETHYST sailed and CONSTANCE commenced her official duties. All was not work however, most watchkeeping duties would revert to day duties, two football teams went ashore and played a match at a ground which bore the rather grand name of Fratton Park. "There was little in common between the pitch and the Fratton Park we all knew" remembers Jack.

Both a footballer and a referee, young though he was, when he arrived back in Hong Kong, he took an examination in refereeing on board H.M.S. BELFAST, the adjudicators being a Commander, a Warrant Officer and a C.P.O., he recalls "Passing comfortably".

Thousands of miles of patrol duties were steamed by ships of the Fleet in Malayan waters, keeping out unwelcome intruders, bombarding to assist the people ashore, and carrying troops police and others who needed to gain access to areas not easily reached by land. Jack Mead remembers spending one night on duty helping a party back aboard, feeding them, finding them drinks and looking after them generally.

Returning to her visit to Nanking, on 19th. December the Frigate ALACRITY took over the guardship duties. CONSTANCE carrying the Czechoslavakian Ambassador down river, ran aground in fog the next day. Her motor-cutter took the Kedge anchor out, lowered it, signalled the bridge that this had been done and CONSTANCE 'kedged herself off the sandbank', resuming her journey.

On 13th. April 1949 CONSTANCE was rostered to go to Nanking again, however her complement was due for a routine chest X-ray and AMETHYST was ordered to carry out the Yangtze duty. On 20th. April reports were received that AMETHYST was under fire in the Yangtze. CONSTANCE slipped and made for the river estuary at 28 knots. Fog intervened and she had to reduce speed, eventually with very low visibility she had to anchor. When fog had reduced, CONSTANCE was ordered to Shanghai, where she made fast alongside LONDON. CONSORT and LONDON had attempted to help AMETHYST escape, and like the frigate had themselves suffered damage and casualties. Warships are in their natural element when at sea, in the restricted confines of a river they are very vulnerable, particularly where well sited shore batteries are involved. It was decided therefore that no further attempts to go up-river would be made. One Short Sunderland of the R.A.F. was able to land and ferry across a medical officer who was of great help to the remainder of AMETHYST'S crew, since not only were wounds to be contended with, but the prospect of disease in the circumstances to be found on board, was ever present.

Several members of CONSTANCE'S company were transferred to the LONDON, and some to CONSORT to replace the killed and wounded.

May arrived and CONSTANCE was again in the Woosung area where the sound of gunfire was a constant reminder of the conflict onshore. many Nationalist soldiers were flooding into Shanghai to bolster the garrison trying to hold off the attackers. On 31st. July CONSTANCE was sailed with orders to support AMETHYST who had radioed that she was attempting to escape down river. Conditions aboard her were such that the only choice open to Kerans, her Commanding officer, was to break out or to evacuate and destroy his ship. Given the means to break out, which he had, it was inevitable that he should decide to go for the chance to escape, risky as the project would be! CONSORT and others also had sailing orders to offer whatever support they could and it was CONSORT which first met AMETHYST off Woosung, the sound of gunfire as the frigate exchanged fire with the shore batteries, preceeding the meeting. The good news of this wonderful escape lifted the hearts of the British people, not least the sailors of the far Eastern Fleet who had looked on with a sense of near despair that one of their own was in such straightened circumstances.

Peter Carlisle from Banbury served aboard CONSTANCE from 1949 to 1951 and he relates how in 1950 the ship was hit by a particularly large wave which put her over to

- it is said - 36° or so. Able Seaman Cripps was washed overboard and remembers finding himself on a Carley Float, he has no idea how that came about. One of the petty officers got him back aboard, and discovered later that two petty officers on deck when the wave struck, were themselves washed over by the sea and into the guard rails. They too were lucky to get to shelter and safety.

In February 1950 a snowstorm in Tokyo Bay brought visibility down to circa 100 yards, CONSTANCE was tied alongside COSSACK at a buoyed mooring. Difficulties arose, CONSTANCE in an attempt to slip from her moorings found them to be fouling COSSACK'S. Some damage was sustained in what was a nasty moment, but eventually with her engines at slow ahead she was able to anchor and ride out the foul weather.

E. Lindsay, a rating from the Light Fleet Carrier THESEUS was knocked overboard when a landing Sea Fury accidently released an 80lb. missile. CONSTANCE was quickly on the scene and both Lieutenant Carne and Surgeon Lieutenant Stan. Walsh dived over to save him and get him aboard. Both men subsequently received an award from the Royal Humane Society.

Wartime service around the Korean coast involved the ship, in common with her sisters in patrols, bombardments, and escorting other units and merchantmen. Around mid-December 1951, CONSTANCE was hit by shore batteries at Amgak, but the damage was not too serious although she sported a hole some two feet above the waterline, starboard side, which would have been uncomfortable enough. Later that month the Daily Express reported: "For 100 days and nights British and other United Nations warships, including the destroyers COSSACK, COMUS, CONSTANCE and CHARITY have been battling in icy weather to hold the islands off the North Korean coast against Red invaders. One British sailor was killed aboard the destroyer COCKADE and two men are missing. The destroyer CONSTANCE was holed near the waterline, starboard side, by shore fire at Amgak''.

Christmas 1951 was spent in Japanese waters. More action was experienced in January and February supporting the forces ashore, inshore at Sokto, CONSTANCE was near-missed, she captured 2 junks in May, and on the 18th. of that month she embarked Rear-Admiral Scott-Moncrieff (F.O.2), who had been Rear Admiral Destroyers at Malta in 1947 when the CH's had formed a significant part of his strength. CONSTANCE delivered the Admiral to the cruiser BELFAST whilst that ship was at Choda. Further Korean work followed, but CONSTANCE went aground, and as a result both her 1st. Lieutenant and her Captain were court-martialled at Hong Kong in August. Commander P. U. Bayly D.S.C. and bar assumed command on 23rd. August 1952.

Bombardments continued, CHARITY accompanied her from time to time, visits to Japan for repairs, replenishment, rest and recuperation were used instead of the longer trip to Hong Kong. After the Korean conflict had ended, exercising, patrolling and all the usual peacetime activities began, in particular the Formosa Patrol was a feature of small ship life since the area of sea between the mainland of China and Formosa, an island some 250 miles long by 110 miles wide, and the seat of the Chinese Nationalist Government, who had by now lost the mainland of China, was often an unpleasant place for shipping, being liable to attack by various factions, including pirates.

CONSTANCE re-commissioned on Sunday the 6th. December 1953. After an introduction by the Captain, the Chaplain addressed the ships company, commencing in the following fashion:- "Brothers, seeing that in the course of our duty we are set in the midst ...." Surely the only serving officer in the Royal Navy to be able to get away with the use of the word Brothers, in the modern trade union context, (So it could be thought), rather that the much more widely accepted biblical context of brotherly love!

In 1954 CONSTANCE returned to the United Kingdom. Forming part of the 3rd. Destroyer Squadron (Destroyers were in squadrons from January 1952). She served in the Mediterranean for 3 months, and in the Autumn was reduced to the reserve at Chatham. Earmarked for disposal in 1955, she finally succumbed to the ignomy of the breakers torch, in March 1956, after a short, but full life. Ward's yard at Inverness was where the dirty deed was done!

♦

Guard of Honour found by H.M.S. CONSTANCE at the funeral of President Hoxa of the Philipines, Manila. Late 1947.

Meanness. Being deaf and not telling your barber.

CHAPLET

H.M.S. CHAPLET was a Portsmouth manned ship, she had an interesting life interspersed with the almost inevitable periods in the reserve fleet and consequent inactivity. Completed at war's end in August 1945 she joined the 14th. DF for service in the mediterranean Sea. Showing the flag was part of each warship's duties in peacetime, and she visited many of the countries in the Mediterranean Basin. The major work was the control of illegal immigration into Palestine (now Israel). Many Jews who had survived the nazi persecution in Europe during World War II now were seeking their own homeland, whether they had ever been there before or not. Palestine was a British Mandated Territory and British forces were maintaining an uneasy peace between Jew and Arab, whilst also trying to maintain the status quo by denying illegal entry.

Access to Palestine by Jews was restricted to the coastline since Arab countries surrounded the Palestine Mandated territories. Many potential immigrants were cheated out of their money by unscrupulous agents in Europe who offered them a passage, the lucky ones found themselves travelling to a port on the Northern shores of the Mediterranean, or Black Sea, where a ship-of-sorts awaited them. In the early days caiques were often used, as time progressed, many larger vessels were used, up to 4000 tons or so and often found by French or American sources. Almost always overcrowded, they would sail a circuitous route in an effort to confuse the searching Royal Naval ships. Radar, aircraft and good intelligence information however tended to give the British an advantage over the illegal transports, and most were arrested.

The main British naval base in the Mediterranean was at Malta, some 1400 miles away, Alexandria and Port Said in Egypt were fairly close, and Haifa in Palestine itself was used extensively for illegal immigration control. Cyprus was in British hands, and Crete was utilised to some extent.

In numerical terms, destroyers and frigates made up the bulk of the Palestine patrol. But other vessels, both smaller and larger were employed on occasions, when necessary. Beirut in the Lebanon, in recent years the scene of some bloody fighting, was used for short periods where the sailors could enjoy some respite from the ever present threat of terrorist action, and from long periods on sea patrol.

Illegal immigrant ships would be shadowed often for days before reaching the 100+ miles of coastline to be defended. Where 3 or more ships approached the shore simultaneously, things became difficult for the warships since great determination was shown by the Jews to get ashore, and, by International law, it is only lawful to board a ship within territorial waters, it is acknowledged some got through and disgorged their immigrants into the waiting arms of Haganah.

To get aboard the ship when signals to stop had been ignored, it was customary for two warships to take station on the two quarters thus steaming behind, and slightly to the left and right of the line of advance of the target ship. Tactics revolved around the amount of resistance the immigrant ship was prepared to offer, this usually took the form of running across the decks in unison thus creating a difficult rolling motion, to the use of small arms as the larger steamers came into use. Food tins and bottles filled with saltwater or urine were hurled across whilst the ship rolled and jinked to avoid boarding.

After several injuries to the boarding party were experienced, by knitting needle or hatpin holding women, cricket box protection was to become standard issue for boarding party ratings. Boots, gaiters, one arm protection as in batsmans wear, one stave, and a paratroopers type helmet painted white, which could be easily spotted from the warships engaged, should a man be surrounded by hostile immigrants. Small arms issue varied according to circumstances, but was often restricted to one or two senior people. Later on however, as the hired thugs became more prevalent and firearms produced, particularly on the large steamships, it became necessary to arm men generally.

Kid glove treatment had to be used - where possible - since the outcry in the world Press may be easily imagined should a Royal Navy ship ram and sink a shipful of Jewish immigrants attempting to reach their homeland, most with concentration camp numbers on their arms! Yet the very methods adopted by the skippers of the blockade runners was quite likely to achieve that desperate result.

Eddie Barrett of Scunthorpe remembers that CHAPLET was involved on stop and arrest work and whilst attempting to board a particularly difficult ship, both tear gas and thunder flashes were used during the action. Some of these were promptly thrown back onto CHAPLET'S deck, thus creating a gaseous atmosphere within her engine room since the fans which supplied the furnaces with air took their suction from the upper deck as did the ventilation fans. Respirators therefore had to be donned in the engine and boiler rooms. The engineer officer of the watch on putting his respirator on, discovered a cockroach in his eyepiece, removing the mask, he received a dose of tear gas to the amusement of the stoker's messdecks later, when the action had subsided.

This arrest led to the Jews, as was usual being taken into Haifa, so near yet so far! Subsequently to be taken under escort to Famagusta in Cyprus where they would be interned in a barbed wire encampment not unlike the ones in Germany, Austria and Poland, except that they knew they were in no danger of ill-treatment or death. The irony of the situation would not escape them, since it was the British who had stood alone in 1940 to protect Europe and such people as these!

Eddie Barrett and Dennis Barker of Leeds recall that a Seafire aircraft had crash-landed on the Italian Isle of Lampedusa. CHAPLET sailed to recover it, a strange job for a destroyer, since she is small with precious little deck space available. Local labour, complete with horse and cart were organised, and the plane plus the unlucky pilot were carted to the ship, a sight which may have won some enterprising photographer the "Picture of the year award" - at least in Lampedusa!

Further ignomy occurred at the jetty when the aircraft was loaded into a fishing boat, locally owned, which was towed by CHAPLET'S motor boat to the ship where she was hoisted aboard and made secure in the ship's waist and the journey to Malta. The "Times of Malta", then still edited by a Britisher, Mabel Strickland, reported the event in an amusing fashion by referring to the "Aircraft Carrier" CHAPLET entering harbour.

Among other duties, the ship acted as escort to one or other of the Light Fleet Carriers then serving in the Mediterranean, OCEAN & TRIUMPH.

Eddie Barrett became the motor boat "stoker" and relates that during a visit to Alexandria, King Farouk was invited to visit the Cruiser LIVERPOOL. LIVERPOOL was to fire a Royal Salute, and it is customary in those circumstances, among many other things, for each boat in the vicinity to stop engines, as a mark of respect for the personage being saluted. As he put it, "The engine was a pig to start," and being in what he thought was a fireproof situation he left the engine running, but out of gear. His fireproofing was insufficient and someone reported him, result? Captains Report.

After the State of Israel had been formed, there was deep unrest in the surrounding Arab States, in addition to that within Israel itself. CHAPLET found herself at Aqaba in Jordon in late 1948 to help, it is thought, the Arab Legion defend itself. The soldiers were short of water and victuals, and they were invited aboard and fed, each man being able to take a much needed shower before returning ashore.

Visiting Istanbul, Dennis Barker remembers each member of the ship's company receiving a box of cigarettes from their Turkish Navy hosts. 200 in each box, which was adorned by the Turkish and the Union Flags. Unfortunately no-one could smoke them, and he remembers cards being played for them. "Have you ever tried losing at cards? - I finished up with hundreds of the bloody things".

CHAPLET sailed home to Britain and went into reserve in 1951.

Later CHAPLET became part of the 1st. Destoyer Squadron, and was converted to a destroyer/minelayer, having one of her 4.5 main armament guns removed to allow this to happen. Destroyers were now in Squadrons, not Flotillas as heretofore.

She returned to Mediterranean waters in 1955 where one of her duties was escorting the Royal Yacht Britannia, Leading Seaman Derek Powell recalls. Returning to home waters she again carried out guardship to the Fleet Aircraft Carrier ARK ROYAL. On one occasion she rescued the crew of a downed aircraft in 10 minutes. The ship carried Royal marines via the Kiel Canal for a NATO exercise in Autumn 1955. Returning to Rosyth at the end of September. Derek recalls a happy ship.

Dave Killelay of Gainsborough joined CHAPLET in the 1950's for his sea training, prior to moving to a more permanent appointment.

Derek recalls a nasty moment off the East coast of Scotland, last in line to six destroyers, pulled out to overtake when the order was received to "Reverse the line", the ship in front also pulled out resulting in a collision. CHAPLET, struck on the quarter, rolled, hung for a while, and then rocked in pendulum fashion, until she once more regained an even keel. No lasting damage resulted.

CHAPLET paid off at Devonport in 1961, and was taken to Blyth in late 1965 for breaking.

◆

Heard in the Elephant & Castle, Jack home on leave, cap flat'a'back, white silk scarf tucked behind his silk, slightly rocky 6 pint movement from toe to heel: "When you hear 'Hands to Dinner' piped in RNB Devonport, you only need to stand at the bottom of the steps in the crowd, no need to walk, you go on in whether you want to or not"......

Horse drawn Seafire, late 1947.

Loading Seafire onto CHAPLET for passage to Malta.

CHAPLET buries her nose in.

Boarding May 1947.
Smoke bombs and thunder flashes evident. ➤

Boarding party appear in control.

H.M.S. CHAPLET, messdeck, Starboard side, aft. Here men would live, eat and sleep, only leaving the mess to work whilst at sea. The horizontal bars near top of picture, with slight downward indentations were where the hammocks would be slung. Getting out for the 4am watch without disturbing your mates either side was an artform.

H.M.S. COSSACK

Six Royal Navy vessels have borne the illustrious name COSSACK. The 5th. a Tribal Class Destroyer, probably best remembered for boarding the German prison ship ALTMARK in Norwegian waters and releasing captured British seamen, had a first class war record until she was torpedoed and sunk by a U-boat later in the war.

The COSSACK we shall consider, the 6th. was completed and ready for service in August/September 1945 when Allied warships were already in Sugami Bay, Japan. United States troops were ashore in Korea (then Japanese held), and the war virtually over. To all intents and purposes it seemed that the peacetime function of the ship would be the usual one of showing the flag, patrolling trade routes, exercising and training for war, and the many other duties which befall a Royal Navy destroyer.

Destined for the 8th; Destroyer Flotilla in the Far East, COSSACK steamed there and became Flotilla Leader. She was to be the first British warship to visit Korea after that countries liberation.

Cyril Allwood served in COSSACK from November 1946 for one year. He recalls being in Hong Kong for Christmas 1946 when he took part in the ships choir which commenced it's round of entertainment when it sang three tunes on each ward of the R.N. Hospital. Next stop was Governors House which at that time was occupied by the Admiral Commanding the British Pacific Fleet. Moving on, both Army and Royal Air Force bases were carolled, whilst refreshments, predominantly liquid it must be said, were taken from time to time.

By 4 a.m. Cyril remembers, their capacity to sing had been reduced. More noticeable to the casual onlooker perhaps, was the deterioration in their ability to sing together and in tune.

This detail in no way restrained their entry into the Hong Kong Hotel, where, led by the officer in charge, complete with drawn sword, they carolled and sang the Russian "Cossack song" as an encore!

Arriving at COSSACK'S berth at 7 a.m. the final attempt was to serenade their own ship's company. This was frustrated by the Captain, who, like Queen Victoria before him 'was not amused' and ordered them aboard.

History does not record what was said later on Boxing Day.

March 1947 found COSSACK at Sasebo in Japan where they were royally entertained by the Americans who had a base there, since the sailors had no access to the necessary dollars, the Americans served them free beer, naturally this proved embarrassing to the sailors who often felt the poor relation when in the company of our cousins from across the sea.

On a warehouse in the dockyard the Americans had erected a very large sign indicating thet: THROUGH THESE PORTALS PASS THE FINEST FIGHTING MEN IN THE WORLD. The temptation to improve on this proved to much for jack, and overnight a further notice appeared noting that: IT IS THOUGHT THAT AMERICANS SOMETIMES ALSO PASS THIS POINT!

After visiting the U.S. Base at Fukuoaka in Southern Japan between 5th. and 8th. April 1947, Captain (D) in his report of proceedings to the Commander in Chief noted: "In Fukuoaka, the coloured U.S. troops are as adept as they are liberal in the use of broken bottle, knife or razor."

COSSACK took part in several acts of assistance to merchant ships who found themselves in difficulties. On Friday 12th. May 1950, whilst patrolling off Amoy, on the Chinese coast Northwest of Hong Kong, and between that coast and the island of Formosa, now Taiwan, she went to the assistance of the ETHEL MOLLER, a ship of some 560 tons which had been captured by the Chinese Nationalists in February 1950.

Captain R. White of Maidstone, COSSACK'S Commanding Officer, ordered the boarding party away, directing them to take over the bridge and engine room first. The 14 strong boarding party took the boat across toward the ETHEL MOLLER where armed soldiers lined the upper deck railings, somebody made the remark "This is not what I trained for as a stoker" which lightened the atmosphere somewhat in the boat, whilst all were waiting to see whether the soldiers would take aim on what would have been an easy target. The fact that COSSACK had trained her main armament on the merchant ship no doubt coloured the thinking of the Chinese!

On boarding, the ship was in Royal Navy hands in a few minutes. The skipper, Captain Waites, and his crew were relieved to see the navy men take over from the soldiers, COSSACK now sent over her Steaming party, whose function was to operate the vessel and get her back to Hong Kong, most of the Boarding Party returned to the destroyer. Amongst the newcomers was the Flotilla Medical Officer, Surgeon Lieutenant R. Morgan, Royal Navy, whose task was to deal with four casualties sustained by the merchant seamen on the previous day whilst under Communist gunfire from shore batteries. Morgan realised that some 150 soldiers were aboard, including 2 generals, a handful of women and children were also found, having treated the wounded who were housed in filthy conditions, he found them suitable accomodation by turfing the generals and their ladies out, and inviting them to fend for themselves.

The military were disarmed by the sailors, and the weapons and ammunition locked away, 3 boxes of silver found on board were kept on the bridge throughout the voyage, under the eye of the senior R.N. and Merchant Navy officers all the way to Hong Kong.
COSSACK returned to her patrol when the frigate WHITESAND BAY took over escort to ETHEL MOLLER. That no casualties were sustained during this operation is quite remarkable, things would have been so different if the soldiers had defended their prize.

Cyril Allwood recalls that each member of a ship's company would receive salvage monies according to rank or rating. In the event of a successful rescue leading to a ships safe passage to harbour. Two payments were received by him which he has still recorded in his service documents, these totalled £7-0-4d.. Insignificant today, but well worth having on navy pay in 1950.

During 1949 he had received less comfortable news when he heard that his cousin, coxswain of H.M.S. CONSORT had been killed at the wheel of his ship whilst she was attempting to render assistance to H.M.S. AMETHYST in the Yangtze-Kiang, after being disabled by Communist guns.

In addition to her many other duties, COSSACK carried out 3 patrols of the Yangtze River estuary in 1950.

During April 1950, she visited Ominato, Honshu, Japan, in company with the cruiser JAMAICA. The town Mayor was entertained on board the two ships, and returned the compliment by inviting the two captains and all available officers to a party ashore, where ladies from his own house of pleasure served crab, tea and saki.

Twenty fifth of June 1950, saw the commencement of the Korean war, when the Communist North Korean forces attacked the South Korean border, thus bringing in most of the Western allies who, through the United Nations, were committed to defend the South. By 8th. July COSSACK had been fired on by shore batteries. She herself was to carry out 171 bombardments of shore targets in support of the allied forces, and during the conflict she expended over 4000 rounds of 4.5 inch and 2400 Bofors rounds. During a refit in Hong Kong in August 1950, the close range 20 millimetre Oerlikons were replaced with the much more effective 40mm type.

Bread was often supplied to the British ships by the large American vessels, as was oil etc. Whilst going alongside a tanker to refuel, the stern of COSSACK brushed the stern of the tanker and a depth charge was dislodged from COSSACK'S racks. Fortunately for both vessels, the primer was set to safe. However both ships moved away from that particular berth as a precautionary step. Activities on the West coast of Korea occupied the ship for most of her time, but occasionally the Eastern seaboard saw her make a contribution. One such duty involved steaming into the bay at Wonsan where Royal Marines occupied some of the islands as a base for raiding operations on the mainland, support would therefore be offered wherever possible.

Alan Quartermain of Banbury recalls COSSACK spotting for train movements on the coastal railway, as a train poked it's nose out of a tunnel, a star shell would be fired to illuminate the target, and high explosive shells from the other 3 guns of the main armament, each previously loaded ready for action would be fired quickly before the next tunnel hiding place was reached by the train. Alan also remembers the mines both moored and the free floating variety which the Communists were prone to use in the sea between Korea and China. He describes how his skipper was fond of taking shots at them from the bridge, to explode and thus render them safe. Even a Bren gun was used for this purpose. Long periods at action stations, and the lesser defence stations are recalled, advice to refrain from using hammocks, and sleep on the decks - where possible - when in possible combat areas.

When the Chinese put their weight behind the North Koreans, even greater vigilance was required since they had considerable air and naval forces, in addition to a massive army. COSSACK claims to have bombarded further North than any other ship, when she operated in the Yalu River estuary. It is noteworthy that the ground forces of the German defenders of occupied France near the Allied bridgeheads found the naval gunfire support particularly destructive, and have suggested that it formed a major part in the Allied success.

The following message was released to the world press in October 1951: H.M.S. COSSACK Senior Officers ship of the 8th. Destroyer Flotilla operating off Korea, Commanding Officer is captain V. C. Begg D.S.C., R.N., has just carried out bombardments along the West Coast as far North as the Yalu River. During bombardment she shelled enemy troops on the march. In past 4 months COSSACK has steamed 22,611 miles in Korean waters, and spent 83 days at sea. She has used 1500 rounds of 4.5 ammunition in dispersing 2 platoons of troops, strafing others in slit trenches, destroying 10 barracks, damaging 10 bridges, railway sidings and yards, rolling stock and 2 gun emplacements. One shore battery silenced, and airfield damaged. Further targets included Wonsan, Songjin and Chongjin on the East Coast.

Shortly before the Korean War, COSSACK was instrumental in landing people at Ominato, Honshu, Japan as a construction force. By towing the ship's bottom line between two jeeps, ground was levelled to furnish the fleet with a summer base, 3 football pitches, 4 hockey pitches, rifle range for 200, 300 and 500 yards, and a 9 hole golf course.

Historical note:- The Boarding Party at the ETHEL MOLLER incident was:

Lt. Cdr. T. G. Ridgeway (1st. Lt.)
Surgeon Lt. R. Morgan.
POSM Fahey J.
L/Seaman Kingsland F.
ERA Knights H.
EM Kermeen J.
SBA Sharp E.
L/Seaman Stevens D.
L/Signalman Wakefield N.

The Steaming Party was:-
Lt. Cdr. H. J. Bartlett
Mech. 3 Bristow C.
AB Haywood B.
AB Luke A. C.
CPO S/M Marsh V.
AB McDowall R.
O/Seaman McGuffie J.
Telegraphist Sawyer E.
O/Seaman Todd A.
O/Seaman Warrener

The following were in both Parties:-
SM Baker A. L.
Elect. Harrap F.
PO Orchard C.
SM Reid D.
AB Smith L. G.

COSSACK was to spend a great deal more time with the 8th. D/S before reaching Devonport in December 1959. On 1st. March 1961 she was taken to Troon for scrapping.

◆

Here comes "Wet paint" Cossack - First Lieutenant's got shares in ICI you Know!

CHEVIOT

Alex. Stephens Yard at Govan laid down the keel of a ship on 27th. April 1943, CHEVIOT began to take shape during a week when there were many portents of events to come. Three days before, two IRA men, Messrs. McAteer and Steele, commandeered the Broadway Cinema in Belfast and made propaganda speeches, later to escape. Lord Haw-Haw, the British born German propagandist was to make much of this event via his broadcasts to the German, and other peoples.

One day before the keel laying, Subhas Chandra Bose who was an ardent Indian Nationalist, very much in favour of co-operation with the Japanese, transferred from a German submarine, in which he had travelled from Germany, to the I29, a Japanese submarine, in the Indian Ocean. Duly carried to Penang on the Malaysian Peninsula, he was put ashore to continue his anti-British work. On the same day, President Roosevelt's personal representative in India had been refused permission to visit Mahatma Gandhi, the imprisoned Moslem leader who also adopted an anti-British stance. The American envoy complained, as did the Moslem League who demanded a separate state to be known as Pakistan "to prevent Bloodshed".

After the war when partition finally occurred there was a tremendous upsurge in violence and bloodshed, and even today (1994) there is still an uneasy peace between the two. Gandhi was released by the British - on medical grounds - On 6th. May 1944.

Launched in May 1944, one month before the Normandy landings, CHEVIOT was completed and commissioned in December 1945 into the 14th. Destroyer Flotilla which as we have seen was to serve largely in the Mediterranean Sea, as part of a peacetime fleet.

Things were not however, as peaceful as the nations sickened by six years of brutal war would have wished. Spain continued to press her demand for the return of the Main British Base in the Western Mediterranean, ownership of Trieste in the Northern Adriatic was in dispute, and terrorist attacks took place from time to time. CHEVIOT took her turn in trying to calm matters down by her presence. Albania at the lower end of the Adriatic was to all intents and purposes a closed book to the outside world, however she had made her presence felt when, in May 1946, shore batteries had opened fire on the British Cruisers ORION and SUPERB as they negotiated the Corfu Channel which at its narrowest point is only 3 miles wide. No hits were scored, but the ships moved into their destination, the harbour at Corfu, where Rear-Admiral Kinahan signalled details of the incident to the C-in-C, Admiral Sir Algernon Willis. British newspapers did not publish this news until more than a week had passed.

On the 17th. October, the Cruisers LEANDER and MAURITIUS in company with the destroyers SAUMAREZ and VOLAGE arrived in Corfu for a four day visit. They left on the appointed day and most men were surprised to hear Action Stations sounded. Many had not even heard of the incident earlier in the year. VOLAGE believed a machine gun was fired, but unsure, did not return fire. Both destroyers were to be severely damaged by mines laid by the Albanians within a short space of time, SAUMAREZ first to be hit, had some assistance from VOLAGE, who then suffered the same fate.

Both ships, though severely damaged were eventually towed to Malta. 44 men died in the two ships.

The situation in Greece was as volatile as it had been during the war, although the Germans had kept the lid on to some extent. The Greek Peoples National Army of Liberation had been formed in 1942 by the communists, fighting took place in Athens at wars-end, and British troops, assisted by R.N. ships acted once again as peace-keepers.

French occupation of Algeria was being contested by factions within Algeria (and France). France was now once more our ally after the split loyalties of the de Gaullist's and the collaborationist Vichy Governments. Egypt was demanding British withdrawal from the Sudan, prior to that country becoming part of Egypt. The ownership of the international waters of the Suez Canal was being questioned, and Egypt was casting covetous eyes on the vitally important strip of water.

By far the greatest demand on fleet resources however, was the work of maintaining an effective patrol off the Palestinian coast to deny entry to illegal immigrants, known to British servicemen in that area as II's. (Eye Eyes).

From almost every country on the Northern Seaboard of the Mediterranean, and from some Black Sea ports, these II's would be gathering with the intention of gaining access to Palestine. CHEVIOT was to play her part in this work.

In February 1949, Bill Brown of Bournemouth joined CHEVIOT at Malta as Acting Yeoman of Signals, as such he would have spent the greater part of his working life on the bridge of the ship, the nerve centre before the more modern operations room became the vogue. Apart from signals staff, officers and lookouts would be stationed there and as a consequence, they knew far more about what was going on than did most of the ships company below decks. Since Bill served aboard CHEVIOT and two of her sister's we will meet him again elsewhere. He states that the ship visited Aqaba in Jordan, and after carrying out guardship duties at Trieste, CHEVIOT was relieved by a United States destroyer.

Bill remembers that whilst Radio Telephony was installed and used in H.M. ships in the early post war years, the more traditional skills of signalling by light, by signal flag hoists, and by semaphore, were still much used.

May 1950 saw the ship's companies of CHEVIOT and CHEVRON change over, the former being sailed home by her new company, repairs being required to her turbines.

COSSACK class ships made up the 8th. Destroyer Squadron in the Far East, however CHEVIOT was ordered to join the CO's for a commission in 1956. Since the ship was at Mombasa in East Africa, her crew flew out in Hermes aircraft, (56 passengers) from Blackbushe, some landing in the South of France, North Africa, Central Africa and finally at Mombasa itself. It seems the Suez war propelled the Admiralty into this method of re-commissioning a warship, British minesweepers undertook the task of clearing the canal of mines after the Suez War.

Countries visited during the commission ranged from Korea and Japan in the North to Australia, and Christmas Island in the South. Captain W. Malins was in command, since CHEVIOT was a leader, and he was later to write, "She has been the finest ship I have ever commanded".

Charles Wardle, an Engine Room Artificer of Burton on Trent served aboard during the 1956-8 commission, and remembers the wide variety of ports visited, from the utter squalor of the more undeveloped areas, to the overwhelming luxury of parts of Hong Kong and Singapore. Proud of the ship's sporting record, (she won the cock of the fleet for boat-pulling) he himself took part in the hockey team. Trincomalee, once a large R.N. base, was at the time of CHEVIOT'S visit reverting to jungle.

Flying servicemen to stations abroad had now become commonplace, and as the tour reached it's end the new company arrived at Singapore by air. CHEVIOT re-commissioned on 5th. May 1958. Working-up was carried out, sometimes in company with the Dutch GRONINGEN. Trincomalee was re-visited as rioting was taking place in troubled Ceylon (Sri Lanka).

After a long refit in Singapore, Saigon was visited, and Hong Kong reached in time for Christmas. 1959 saw her acting as planeguard to the Carrier ALBION, and in company with H.M. Ships CEYLON, CAVALIER and CHICHESTER, and the Commonwealth ships QUEENBORO' and QUIBERON, she escorted the Royal Yacht BRITANNIA from the Mallacca Straights to Australia, visiting Albany and Freemantle.

Whilst refuelling from the tanker OLNA in poor weather conditions CHEVIOT was struck by a large wave which swept a man overboard. CHICHESTER who was in company recovered the seaman whose name was Galliford, unconscious by this time. He made a complete recovery.

During the period 1948 to 1960, a state of emergency existed in the Malayan States, CHEVIOT and most other R.N. ships which served in the area during that time would carry out patrols, investigate ships and boats etc. in aid of the land-based forces and the civil power. Large scale exercises with other members of SEATO were also a feature of the work carried out by ships of the Far Eastern Fleet.

March 1960 and CHEVIOT was based at Rosyth where she formed part of the training arrangements, she relieved H.M.S. TALYBONT.

◆

*Moment of boarding, white helmeted Boarding Party.
Note scaffolding and wire netting on destroyer to intercept missiles from illegals.*

CHEVRON

H.M.S. CHEVRON was the first of the CH Class to be completed for service, coinciding with the end of World War 2. She underwent stability trials in addition to all the other tests such as main machinery, speed trials, gunnery performance and the rest.

Mr. W. Bartle of Longriddy, East Lothian was amongst the first of her company to join her in Alex. Stephen's yard, Govan on 2nd. May 1945. On the day that Stalin was claiming to have captured Berlin, and the R.A.F. claimed to have sunk 23 out of 60 U-Boats fleeing from Germany to occupied Norway, a total reversal of fortunes from the grim days of 1942 when the U-Boats came close to bringing us to our knees. Seven days later the rest of the ships company joined CHEVRON. Ordered to the Mediterranean, she began the rounds of showing the flag, exercises, refits and the Palestine Patrol. When King George the Second of Greece died, H.M.S. CHEVRON provided a guard of honour at his funeral in Athens in 1947. Entirely appropriate since the Greek Royal Family had a close affinity to the Royal Navy, visiting British warships from time to time. Many who served in the R.N. during those early post-war years will remember the vivacious Queen Frederica being greeted at the gangway by the Captain (or Flag Officer where carried).

Ghain Tuffieha, (pronounced Hine Toughear) in Malta's North-West corner provided the training ground for R.N. Officers and men selected for boarding party work - this training proved vital when the leap from warship to illegal immigrant ship was made. Broken ankles and other injuries were not unknown during this short period of gentle persuasion at the hands of Royal Marine sergeants. Nowadays, holidaymakers sun themselves at Golden Bay and other adjacent venues.

CHEVRON was sailing in the Dodecanese Islands when she visited one believed to be uninhabited, and picked up many Jewish illegal immigrants who said that their ship had come to grief, and they had found themselves ashore without sustenance. After carrying them to a British base and handing them over to the authorities, she found that her after mess-decks were lousy with crabs. CHEVRON made for Piraeus, the port of Athens, the victims were found accommodation ashore, whilst the ship was de-loused.

Trieste, long a bone of contention, was offered to Italy at the end of World War 2, but as in so many cases, National sensibilities would not let the matter rest, and the British, and occasionally the U.S. would station a vessel in the Trieste, Venice and Pola triangle. CHEVRON took her turn at this peacekeeping task.

Considerable time was spent on patrolling in the Eastern Mediterranean and CHEVRON was involved in stopping and arresting the immigrant ship ABRA which she boarded and escorted into Haifa where the military would take over and arrange for their onward transmission to Cyprus. Mr. Bartle remembers that Haifa was uncomfortable. Terrorism was mounting, leave, when given often would amount to the afternoon period only, small arms had to be carried. Barbed wire surrounded the dock area, the ship would berth stern-to in order that some 30 yards of water were left between the stern of the ship and the jetty. Charges would explode at intervals, dropped by a motor boat circling the waters of the harbour, and, more inconveniently, thrown over the side by the Officer of the Watch during the night, when they produced an almighty clang on the messdecks. These measures were necessary to discourage terrorist frogmen some of whom had been trained by the British.

The ship took part in the Fleet Regatta and being successful returned to Malta as cock of the fleet.

Bill Brown of Bournemouth served aboard CHEVRON, CHEQUERS and CHEVIOT during service based on Malta. Whilst aboard CHEVRON on a night exercise with the destroyer flotilla, all ships blacked out, steaming in line abreast and at speed, the controlling ship made a signal ordering all ships to turn 90 degrees to Starboard, which would have resulted in a line astern formation. CHEVRON had a faulty R/T set and did not receive the signal, keeping straight on in the darkness the inevitable result was a collision. COMET another destroyer of the CO Class got in the way resulting in damage to CHEVRON'S bullring (The large oval fairlead at the extreme fore-end of the ship). COMET was damaged in the area of No. 2 Boiler Room. It should be remembered that ships showing no lights, steaming in close company with others, at speed, preparing for war conditions, are, from time to time, likely to encounter this sort of misadventure. Royal Navy night encounter exercises provide vital training and not all navies are prepared to fight at night! CHEVRON returned to Britain in 1950.

Commissioned again in 1952 CHEVRON was to be in service over several years. Christmas 1953. Location Sliema front, Manuel Spiteri remembers enjoying a stroll along his native Sliema seafront and observed 'fireworks' from one of H.M. Ships. (The Maltese have a passion for fireworks). He remembers that he had never seen such a display before, now, he realised someone was swimming in the water and when one of the verey lights arced towards where he was, he decided to wander into his favourite cafe, where it would be safer. Meeting a boatman later, he was given a list of ship's names - CHEVRON included - which had taken part. Manuel remembers he used the word 'bang' many times with much arm waving.

1957 to 1964 was spent in operational reserve at Rosyth. CHEVRON was broken up commencing December 1969 at Inverness.

◆

Large caique about to be boarded.
Ship on left is H.M.S. CHILDERS. R91 later, D91.

CAESAR

When CAESAR was laid down on Saturday, 3rd. April 1943, at John Brown's shipyard on the Clyde, Eastbourne was bombed and machine-gunned by the Luftwaffe. Launched on February 14th. 1944 at which time U.S. and New Zealand forces were ashore on Green Islands in the South Pacific, and within the week H.M.S. PENELOPE, a small cruiser known throughout the fleet as H.M.S. Pepperpot, due to the myriad shell and splinter holes evident on her hull and superstructure, was sunk off Anzio in Italy by torpedo from U-410. Thus passed another R.N. legend.

CAESAR served with the Home Fleet as Captain (D) of the 6th. Destroyer Flotilla, Captain Godfrey Brewer R.N. in command. Known to all and sundry below decks as "Slewer Brewer".

During November 1944 she formed part of the escort to the North Russian convoy JW62, and in December of the same year was escorting the return convoy, she finally anchored in the fleet anchorage at Scapa Flow. George Taylor of Chessington served aboard CAESAR for 2 years, 1944 to 1946. He reckons that "Out of the 4 ships I served in, CAESAR was the happiest of them all". George is a member of the "North Russia Club", as is Bill Cairns of Morpeth, who was also in CAESAR.

Their ship was in the protective screen of the convoy when the sister ship CASSANDRA was torpedoed with heavy loss of life early in the morning of an Arctic day, 11th. December 1944.

The following day saw an attempted attack against her when two torpedoes were seen to miss, in addition to which she suffered two minor air attacks. Bill Cairns of Morpeth served aboard during this time as a telegraphist trained to receive morse signals in a variety of specialist languages, Russian included.

On the last day of 1944 she sailed from Liverpool as escort to the large liner MAURETANIA and she left her charge on New Years Day 1945. She went to meet the QUEEN MARY on 6th. January, met her at 12 degrees West, and they both made 28 knots until Liverpool was reached. On 14th. January she met and escorted R.M.S QUEEN ELIZABETH another of the great ocean liners, this time the Clyde was her destination, and both ships made good 28 knots. Six days later both ships sailed the Clyde and again made off at high speed, this could not be maintained by CAESAR however due to foul weather conditions.

It is noteworthy that the great ocean liners relied very much on their high speed performance to keep them out of trouble, and a U-boat would have to be in a most favourable position to put in an attack.

Further escort duties were carried out with the MAURETANIA and AQUITANIA, and CAESAR saw a great deal of the Clyde and Mersey.

In need of a refit she arrived at Devonport where she remained in dockyard hands until the end of May 1945. On the anniversary of D-Day, 6th. June 1945 she escorted the Cruiser JAMAICA carrying King George the Sixth, and Queen Elizabeth to the Channel Islands.

Working up to an efficient unit was carried out, the Clyde was visited, this time to store ship, and after returning again to Devonport, she sailed for the Far East Station on

16th. July 1945. Ports normally visited on passage duly received CAESAR and additionally the Cocos Islands were visited on the 14th. August. H.M.S. CAESAR was at sea on VJ-Day and when the signal had been received, "Splice the Mainbrace" was piped. Clearly a very happy day!

So the war was over! Ah but wait a minute, not for CAESAR it wasn't, as we shall see later. CAESAR visited the British beachhead in Malaya, Singapore and several other areas until she was ordered to the Dutch East Indies where dissident forces were resisting efforts to reinstate the Government after the Japanese had surrendered. CAESAR carried out bombardments at Sourabaya in company with several other destroyers. George Taylor was landed to observe the fall of shot from his ship, after the firing ceased, the ship left for Singapore and George with his shipmates found themselves with the British and surrendered Japanese troops, both of whom found them food and water. He remembers the strange feeling when Japanese who three weeks before had been his bitter enemy, now were preparing rice for him to eat, and clearly were pleased to do so since their treatment at the hands of the insurgents would have been a single bullet through the head.

On a lighter note, whilst in Trincomalee, Ceylon, (Now Sri Lanka), Captain Brewer signalled the Cruiser CLEOPATRA, "CAESAR intends to visit CLEOPATRA after a lapse of 2000 years". He was rowed across by some of his officers, dressed as Caesar, complete with laurel wreath and toga.

At one stage during the war, as a Commander, he was in the Trade Division of the Admiralty, as Convoy Planning Officer. He profoundly disagreed with advice which Professor Lindemann gave to Winston Churchill on methods of protecting convoys. (Lindemann was Churchill's scientific Adviser). Brewer was overheard to say that he "Used to pray each night that Lindemann would be run over by a bus."

CAESAR returned to Devonport in May 1946 and was to spend a long period in reserve.

During the 1950's CAESAR spent some time with the Reserve Fleet, being stationed both in Scotland and Wales. From 1955 to 1956 she underwent a long period of refitting and mothballing, and was laid up in Penarth, South Wales.

1960 came and she was once more needed for service. Modernised, she commissioned for service at home based on Portsmouth. Warmer climes called and she sailed to the Far East, her base for the moment being Singapore. Borneo was the scene of a conflict which lasted several years and CAESAR was to play her part by patrolling that area against pirates and illegal arms shipments intended for that war which was about to commence. Many ports were visited in the Far Eastern area before she finally pointed her stem Westwards and home in June 1965.

She was sold in December of the next year. She carried the Battle Honour 'Arctic 1944'. 1967 arrived and CAESAR was broken up.

Escort commander in destroyer to corvette who had been detached astern of convoy: "What took you so long"? Corvette: "It's been uphill all the way".

CONSORT

H.M.S. CONSORT was launched in October 1944, four months after the Allied invasion of Europe had commenced, and things were going well, despite the fact that Britain, or more particularly the South-eastern portion, was under attack from Hitler's so called secret weapons, the 'V' bombs and rockets. CONSORT was built at Govan by Alex. Stevens, the earlier disruption from air raids had now virtually ceased, and most shipyards could work normally, that's to say within the still severe shortage of most raw materials, including steel.

Ron Morris of Worthing joined her at Govan in 1945 whilst she was fitting out, and was to serve aboard until 1949.

Commissioned for service on the 7th. March 1946, she sailed from Plymouth destined for the Far East Station. Very bad weather was experienced between Plymouth and Gibraltar, Ron remembers heavy steel ready-use lockers being stove in such that the front of some were touching the back. (Particularly bad weather conditions existed in early 1947 in the Western Mediterranean area also). Ron also remembers that various items were lost overboard, but mysteriously, the Chief Petty Officers Mess had won a side of bacon before Gibraltar was reached!

Malta was her base for working up exercises, after completion of which she moved on in stages to her chosen station.

Patrols were carried out around the Malayan Coast where illegal items were carried in by boats, usually at night. This had a plus side since during day-time the ship's company could swim, sometimes to a nearby beach, where the lost sleep could be made up, in between refreshing trips in the warm, salty water.

Britain was one of the occupying powers in Japan, CONSORT had to take her turn at visiting ports, both to ensure that Allied peace terms were being observed, and to offer the crew the recreational activities denied to them whilst aboard ship. Kure and Sasebo were much visited, the Americans had a large presence in the area, and other Allied and Commonwealth powers had forces based at a variety of venues.

Nanking, a large city some 200 miles from the sea was an important shipping and industrial centre. Situated on the Yangtze Kiang River and the Capital City of China, the Royal Navy had a warship stationed there to help sustain the British Embassy, a further duty was to carry diplomats serving a variety of nations to and from the city. Shanghai on the Chinese Coast was the terminal for these river duties.

As CONSORT steamed slowly into Nanking for a further period of guardship duties, authorised so to do under the Treaty of Peace, Friendship and Commerce, between Britain and China, her ships company were aware that the civil war between the Communist North and the Nationalist South was running in favour of the North, and subsequently was bringing the front line nearer to the Yangtze River.

Ron Morris recalls "We were secured alongside, an unused egg-packing station was close by". Any recreation was on a self-help basis, but beer they themselves had shipped up, was available. In the engine room fuel supplies were beginning to give cause for concern with one furnace being constantly flashed-up ready for immediate duty. The tanks were therefore gradually becoming emptier by the hour. Ron goes on to say: "Since I was shortly due to be relieved to return to the U.K., I was working in the gunner's store, prior to handing over to the relieving ordnance artificer, watching empty beer crates being thrown overboard, I asked someone what was going on, "Amethyst is coming up-river to take over and the Chink Commos have fired at her", was the answer".

CONSORT quickly sailed, ignoring the speed restriction, (some 7 to 8 knots it is thought), the engine room staff later stated that at times 27 knots was achieved, this in a river!

"It's alright!" Ron was told a little later, "She's been hit and aground, our turn next" came the same cheerless news from somewhere along the passageway. Before CONSORT was to reach the scene, H.M.S. AMETHYST a frigate, was struck by 53 shells, out of the 180 souls aboard 31 were wounded, and a further 23 were dead, or would die later.

As CONSORT approached AMETHYST'S position she came under fire, which she returned. Soon, still at high speed she saw AMETHYST and realised that the frigate was badly hurt. Passing AMETHYST, she steamed on for a while, her Captain, Commander Robertson who himself was already wounded along with several other men, trying to assess the chances of getting a tow aboard the frigate whilst under destructive shellfire, signalled: "Is it possible to tow you off" AMETHYST replied: "Only if the batteries are silenced first" CONSORT, still firing made her final turn toward the sea, still over 100 miles away and sailed on, her guns were heard by AMETHYST'S crew from time to time as fresh batteries of artillery opened up, down river.

The cruiser LONDON in company with another frigate BLACK SWAN, were ordered up river to assist, however, themselves sustaining damage and casualties, they were recalled.

CONSORT berthed at Shanghai, still in friendly Nationalist China, where she made good some damage, and buried her 8 dead sailors at the Hungjoa cemetary. They were afforded full naval honours at the ceremony.

During a passage between Formosa and the Chinese mainland in the summer of 1951, a rating who was being held in custody, whilst having an exercise period on the upper deck decided to jump overboard. Since the nearest land was estimated by some to be 40 miles away, it is not absolutely certain whether he just fancied a swim, or perhaps the call of freedom was strong enough to drive him on to an epic escape, "Away Seaboats Crew" was piped. CONSORT altered course to effect the rescue, (or recapture) of the miscreant, the Cox'n was at the wheel as befitted the nature of the event. When the seaboat came up to the man, he was offered the boathook to assist him get alongside, but swam to get away, only a sharp crack to the head subdued him enough to be dragged aboard the boat. Having been returned aboard CONSORT and locked away again, the final word was left to the Cox'n "Why didn't you let him go? It'd mean less paperwork".

During her service in Far Eastern waters she formed part of BRITANNIA'S escort, several members of her company sailed an MFV, Motor Fishing Vessel 715, up part of the Malayan West Coast. CONSORT had an unfortunate collision with COSSACK. During the Korean war she took part in every type of activity associated with supporting armies (and aircraft) fighting ashore, supporting aircraft carriers at sea was also carried out. In late 1951, several volunteers elected to serve ashore for a while in Malaya, and were attached to 45 Commando Royal Marines, whilst doing so. They may be seen around November 11th. each year with the red and white General Service Medal with the "Malayan Clasp".

Re-commissioning in 1955 she again served in the Singapore, Hong Kong, Japan areas. Visiting Australia in 1956, after which, a long refit in Singapore, she then made her way across the Pacific Ocean, through the Panama Canal, to Bermuda which she reached on February the 8th. 1957.

H.M.S. CONSORT met her end in a breakers yard at Swansea in March 1961, leaving 16 surviving 'C' Class ships in being.

◆

COCKADE towing off the grounded Dutch destroyer EVERTSEN. Korean war.

Successful man: One who can earn more than his wife can spend.

'EXODUS' near coast of Palestine. Illegal immigrant ship, moment of boarding. Smoke and tear gas grenades in use against resistance, which was substantial. Each side took casualties.

Boarding Party, CHAPLET. November 1947.

Haganah ship 'EXODUS' 1947, damaged by R.N. warship whilst being boarded. Ship was seriously overcrowded, and only passengers with a medical certificate could live in a lifeboat.

*'EXODUS' immigrants disembarking at Haifa. 1947.
1500 Illegal immigrants aboard.*

CHEQUERS. Captain of the Guard greets Queen Frederica of Greece, prior to escorting her aboard, there to be met by the Captain.

Queen Frederica of Greece leaves CHEQUERS after visiting Captain (D).

CHEQUERS in rough weather. 'B' Gun well manned!

Calmer times, CHEQUERS and CHAPLET alongside in the South of France. (CHAPLET, R52)

CHEQUERS

Launched at Scotts Shipbuilding & Engineering Company's yard at Greenock on Monday the 30th. October 1944 by Lady Somers-Hunter, H.M.S. CHAMPION, renamed CHEQUERS was fitted out as a Flotilla Leader.

Belgium and Greece were already free of the Axis tyranny, and exactly one week later, on 6th. November, the Resident British Minister in the Middle East, Lord Moyne, was shot dead by the Stern Gang, the Zionist terrorist movement dedicated to finding a Jewish Homeland. This act in Cairo has a connection with CHEQUERS which will become obvious later in this book.

The CH's were Mediterranean bound, and on the First day of February 1946 CHEQUERS, Captain J. H. Rucke-Keene, OBE, RN in command and designated D.1. with responsibility for the day to day operations of the flotilla under his command, was busy in the Eastern area of the Mediterranean Sea. For four months, away from the main fleet base at Malta, CHEQUERS was to spend most of the time on the Palestine Patrol, set up to intercept illegal immigrant ships carrying Jewish refugees from Europe, intent on landing on the shores of Palestine.

Britain was mandated to administer that country, and immigration into it was strictly controlled. Attempted illegal entry became a growing business, and a great deal of effort was required by the British, each service was involved, as was the British manned Palestine Police.

CHEQUERS helped stop and arrest 3 ships during this period, namely the ASYA between the 26th. and 28th. March, and SMYRNI and JERICO between 12th. and 14th. May. All illegal immigrants (known as II's to the British), were taken into Haifa where the army would arrange disembarkation, prior to movement to Famagusta in Cyprus for internment in the British prison camps.

CHEQUERS was at sea and in company with another destroyer the JERVIS when the SMYRNI was intercepted. Well South of the Palestine coast, SMYRNI was carrying out the ploy of jinking about in order to deny the British the opportunity of boarding her. This dangerous manoeuvre carried on for some time, always carrying the ships closer to the coast. When the SMYRNI cut across the bows of JERVIS, CHEQUERS put a short burst of 20mm Oerlikon across her bows at which point discretion became the better part of valour and she began responding to signals from the two Royal Navy ships. Passage into Haifa followed and the inevitable journey from the homeland to Cyprus began.

In October 1946 the ALMA was intercepted off Haifa, and again a successful operation was carried out and many more II's were transhipped to detention, where some of them eventually would be allowed legal entry to Palestine.

Ian Wardle recalls some of these events, having served aboard CHEQUERS for 2 commissions:- On arrival in Grand Harbour, Malta, aboard the trooper SAMARIA, he, and 20 others were carried to the dockyard where CHEQUERS was completing a 'Care & Maintenance' period, and by then looking distinctly untidy. As the launch approached Ian's new home, groans were heard from some of the men on her upper deck, "Gawd, more bodies, and their ain't enough billets and lockers as it is" this, Ian recalls, was absolutely true. Joining routine he remembers occupied "Two minutes in the Cox'n's office".

Due to the Palestine situation the ship had almost a wartime complement (possibly 30 additional men over and above standard). many therefore had no slinging billet for a hammock, and no locker for their kit. Living out of a kitbag and suitcase was necessary, as was sleeping on a mess-table, camp-beds were pure luxury, some men had only the deck to rest on. Inevitably all facilities were strained, and the bathroom was no exception. Nine tip-up wash hand bowls and one salt water shower for over 100 men, working in a warm, sometimes hot climate, left a lot to be desired.

Clothes were washed and ironed by hand, apart from normal duty and work, it fell to each to take his turn as 'Cook of the Mess' this occurred about every eight days. Two men would be detailed to prepare the food for each meal within the 24 hour period of their duty. Canteen messing required that each mess purchased it's own food from the allowance the Admiralty decided was proper, prepared it, delivered it to the galley at the correct time where the trained chef's would cook it. At the appropriate moment, the mess-cooks would collect it, return it to the mess, having previously laid the table ready, and 'dish it up' to the assembled throng who, having just imbibed their tot of rum would be ready for it, (at dinner-time that is) after the meal, scraps of 'pussers hard' - (soap taken from the bathroom or anywhere else) would be put in a cocoa tin punched with holes, swilled around in hot water placed in a large mess kettle and the two mess-cooks could wash up. On completion, the contents of the kettle - gash - was dropped overboard via a stern mounted chute. An insidious deafness overcame ratings carrying out this operation since they claimed never to have heard that quite distinctive tinkle which indicated that something had gone down the chute other than waste, 'knife, fork or spoon'? never! The service had an oft repeated ditty used at such events, and printed in most books on the navy and the sea, but NOT THIS ONE.

Breakfast often consisted of toast, made on the small electric heater mounted vertically on the round steel column which passed from the lower reaches of the ship up to it's gun mounting above, and which carried the shell and it's case upward on a journey from magazine to breech.

Ian Wardle remembers setting off for the Eastern area of the ship's work, the Palestine Patrol:- "Leaving the delights of Malta with the bull-ring pointing in the general direction of the Levant, boarding party practice was carried out. We had wondered why the focsle awning stanchions had been left in position, and soon found out, rolls of chicken wire were produced from the tiller flat and rigged in place of the canvas awning. From the iron deck upwards, scaffolding poles and boards were rigged, forming platforms for the boarding party to jump from, at one point, these reached to 'B' gun level.

Experienced members of the boarding party drew Lanchesters with magazine pouches, some senior rates had Smith & Wesson revolvers, all wore boots and gaiters, and white, round, steel helmets as issued to the airborne troops, thus giving them some leather protection to the neck. Those without firearms carried staves, wooden, with a bulbous bottom and said to contain lead, but nobody seemed sure!

Two days into the first patrol and an RAF Lancaster reported by radio-telephone a 'possible' illegal immigrant ship. Course is altered, engine room shaft revolutions go on, down goes our stern, 293 Seaguard radar picks up our target, whilst our boarding party musters on the iron deck. Lines, fenders and grappling irons appear, both port and starboard flagdeck

Oerlikons are manned and armed, as was the single pom-pom mounting aft of the funnel. Mess killicks customarily detailed off the mess-cooks to stand by with a steaming bag ready to collect food used as ammunition by the II's which might prove suitable for the mess, and, most important this, would thereby save money out of 'Their Lordships' allowance.

Suspect ship in visual contact, Bridge announces that it will be a single ship boarding as no chummy ship is available. No tug is available to assist either. Speed reduced and now within loud hailer distance. Boarding party divided and secreted in port and starboard passages in break of fo'c'le. Now 2 cables from target, nobody on deck and only two people seen on bridge, CHEQUERS loud hailer asks 'What ship - what owners - where bound'? Now at half a cable, port side to, answer through a megaphone, in a strong American - probably Boston - accent, garbled name, no mention of owners, 'destination Port Said'. Bridge loud hailer asks further questions - 'What cargo - what nationality - do you have radio - I repeat, what ship are you'? Megaphone answers once more - 'Am in international waters with commercial cargo bound for Port Said from Greek Islands'? Conflab on bridge of CHEQUERS, followed by:- 'Correct heading for Port Said is so-and-so magnetic - safe trip'. Still no sign of life other than the two people in the wheelhouse, no flag flying. Clearly CHEQUERS skipper is not happy, but has no alternative but to sheer off.

On go the engine revolutions, the familiar dipping of the stern as we pick up speed, off we go toward North, and over the horizon marking final bearing on Plan Position Indicator, as echo slides off screen. Speeding now near maximum revolutions, swing in a wide arc to port, through West and back onto last PPI bearing. Echo begins to slide onto screen, Chief Yeoman of Signals, his eye glued to his long telescope reports visual sighting 'Dead ahead' and a few minutes later, 'Look at them'. Many pairs of binoculars are raised toward the target ship, people are packed on the upper deck, bodies wherever human beings can climb or wriggle, the ship is classed as a caique of 250 tons.

CHEQUERS had carried out the standard drill, locate, interrogate, pass over the horizon and sweep back to catch them with their guard down. The preparations for boarding were carried out once more. 'Port side to' is confirmed. Fender party position long fenders on port side from anchor to break in fo'c'sle, and shorter types from break to propellor guards. Boarding party emerges from cover, takes position on fo'c'sle near fender party, guard rails removed at take off point, now only 5 or 6 yards separate the vessels, tins are thrown from the II ship, when bottles land on the warships deck and break, the splintering glass may produce nasty casualties. One firecracker is tossed onto the caique, and the effect on those near is immediate, stunned, they turn away, at the order 'go' - the ships are together now - the Club swinger goes first, followed quickly by the rest of the boarding party, truncheons at the ready. Several Lanchester armed ratings cover from CHEQUERS. Wheelhouse and Engine room - first targets - are taken, slight unpleasantness in wheelhouse where one round of Smith & Wesson discharged through scuttle. Some bruising suffered by several members of boarding party, but old hands confirm this was an easy one.

Sea calm, vessels remain secured together until agreement is reached for caique to steam for Haifa under her own power, since Naval Officer in Charge, Haifa, (NOIC Haifa) has advised that tug not available for 3 hours. Mount Carmel to the South-east looking much closer now. CHEQUERS upper deck washed down to remove smell of urine received via bottles from II ship earlier. Boarding party have a further nasty smell under their noses from a small hold forward, which proved to be a desperately poor jury toilet. Destroyers canteen out of 'nutty', some members of ship's company bought it all up and gave it to Il's!

Vessels part and proceed to Haifa at the speed of the caique, late PM we anchor and post anti frogman watch, both motor boats piped away to recover boarding party, messes listing canned food received in the quite unconventional fashion - by air mail - one wag noted. Mess caterer has already calculated savings made to mess funds, boarding party returns home and reports that the army, men of the 6th. Airborne Division it was thought, had cleared the Illegal immigrant ship in 20 minutes, and had shepherded them into the DDT area immediately for decontamination.

No shore leave granted, ship sailed next day at 0500 for a further patrol of 4 days duration''.

As time progressed, larger ships were put into use for illegal work, best known to the world at large was the EXODUS previously registered as the PRESIDENT WARFIELD. Much propaganda resulted from this voyage to the benefit of Jewish extremists, and the detriment of Britain and the Royal Navy. Of 1814 gross tons, she was used as a liner plying up and down the American coast, during the war she was taken over by the U.S. Government, and when the conflict ended was sold as scrap.

In July 1947, she sailed from Sète, in France with some 4500 souls on board. After arrest in Palestine waters, it was decided to return them all to France, there to be offloaded. This, it was hoped would give the French and other like minded governments food for thought (There was much connivance by them, and by the Americans, in addition to the communist states who, whilst ridding themselves of troublemakers, could give the British a headache too - some of our former allies were not shown in a good light at this time!) Given a fruitless two-way journey, paid for largely by the Il's themselves would also teach a salutary lesson it was thought.

Escorted from the Eastern to the Western end of the Mediterranean Sea by R.N. ships, the 3 British transports (captured German ships taken as war reparations) were shepherded alongside at Port du Bouc. Much activity, including meetings with French officials took place in an effort to persuade the Il's to go ashore, without result. They were determined to stay aboard. British officials approaching the vessel on the dockside were pelted with potatoes, and it became clear that an impasse had been reached.

Britain decided that they would then return the immigrants to their original starting point - Germany. Three transports with 3 R.N. ships escorting them, made for Gibraltar. The transports berthed alongside the detached wharf, thus ensuring that any escapee would need to swim for it should the call of freedom be strong enough. Finally each sailed for Hamburg where the British Army persuaded them ashore. Promptly locked them up in our internment camps, it is an irony that Jews with the concentration camp numbers tattooed on their left forearms should suffer the further indignity of once more being forced into camps in Germany - also very sad by any standards.

PAN CRESCENT was the largest blockade runner used, of some 4600 tons, PAN YORK, almost identical was slightly smaller. Leaving Burgas in Bulgaria in December 1947, they carried some 15000 migrants between them. Royal Navy intelligence knew of the proposed attempt well before the due date for sailing arrived, the problems of stopping, and arresting, two large ships packed with a massive human cargo were enormous. Danger of accidents, ramming, or the use of firearms (criminal elements were now involved in the work of transporting Il's, many of whom were robbed, some desperados were known to want a shooting war to occur, presumably to enhance the chances of gaining the shore of Palestine, and secondly to give publicity to their cause).

One cruiser had already been fitted out to assist the smaller warships whilst boarding, however, when the two opposing factions finally met, the Senior Officer Royal Navy present imposed upon the two skippers to sail with him under escort directly to Cyprus, there they were to join most of their predecessors.

Ian Wardle was landed at Limassol in February 1948, carried to Nicosia where the B.M.H. carried out an operation for appendicitis. On recovery, CHEQUERS was not in the area and he served aboard her sister, CHILDERS, and once again found himself on Palestine Patrol duties. In April he was reunited with CHEQUERS. In 21 months of duty CHEQUERS had played a part in the tracking and arrest of 9 illegal immigrant ships.

Israel, formerly Palestine became a state in 1948 and a huge job of work ended for the Royal Navy. Peacetime exercises and training would now go on apace. During 1949 H.R.H. Lieutenant The Prince Philip R.N. joined CHEQUERS as First Lieutenant, much brushing up of the Greek language was carried out as rumour said that much time would be spent in Greek waters, this proved to be false however and the ship behaved in exactly the way the Admiralty would have wished in a destroyer of the Fleet. Princess Elizabeth stayed with the Mountbattens at the Villa Guardamangia in Pieta, Malta, and was able therefore to see more of her husband than would otherwise be the case.

1952 saw the removal of 'X' gun from the ship (The third one from the front or the sharp end). Two Squid ahead firing anti-submarine missile launchers were fitted in place of the gun. Since it was known that Marshal Tito of Yugoslavia would be steaming near Malta in his yacht, (it had not always been his I fear, previous owner:- The Yugoslavian Royal Family). The C-in-C Mediterranean, Lord Louis Mountbatten decided that he would sally forth and greet him in true navy fashion. Tito it should be said had distanced himself somewhat from Communist Russia and clearly did not wish to be dictated too by that State. Further he was a man who clearly enjoyed the advantages that his presidential lifestyle offered him.

On the appointed day, Lord Louis arrived in Sliema Creek, boarded CHEQUERS with due ceremony, watched as his C-in-C's shiny blue barge was hoisted inboard, and was finally escorted below for the trip. H.M.S. CHEVRON another sister ship followed out of Sliema to assist in the calibration of the new A/S projectiles. Lord Louis on CHEQUERS bridge heard confirmation that the yacht was in sight. He ordered that each destroyer would steam directly toward the oncoming yacht, pass her one on each beam, saluting Tito as protocol determined.

This manoeuvre satisfactorily carried out, the two warships soon well astern of Tito, carried out a wide swing, took up station abeam of each other, and prepared, on Lord Louis's orders to demonstrate the new Squid system, on receipt of the executive signal, they were to fire 3 projectiles set for medium depth. The yacht was signalled to the effect that the warships would, whilst overtaking her, fire 3 missiles each. History does not recall what signalling method was used at the time, flag-hoist, semaphore, signal projector light, radio telegraphy (morse) or radio telephony (voice), ''Executive signal sir'' the Yeoman on the bridge of CHEVRON called out, ''Fire'' ordered the Torpedo and Anti-submarine Officer, whereupon his 3 mortars arced into the air at the same time as those from CHEQUERS, (Capt. D1., with the C-in-C with him on the bridge). Something wrong here! CHEVRON'S charges explode under water, CHEQUERS charges explode on the surface of the sea. Consternation on her bridge, the air is blue, large spouts of water containing debris from the 3 exploding charges drop fairly close to Tito. Quickly the T. & A/S Warfare Officer is summoned to the presence and the air is blue for a second time. Radio signals of an urgent nature passed to Tito in apology, and the two destroyers set course for Malta.

CHEQUERS enters Sliema Creek first, Ian Wardle recalls:- "CHEQUERS stops with Creek entrance at Green 90, pointing her stern between the two mooring trots, she steams stern first at more than the usual speed toward the buoys. Funnel gases swirl over the bridge, some are coughing, Chief Yeoman of Signals bellows 'Man overboard'. Eyes look beyond the Starboard Bow and there he is. Immediately the Pilot announces calmly 'C-in-C's barge, bow section overturned Starboard side', another voice speaks of two men in the water, and stern part of barge also adrift in the sea, these two items no longer joined in the middle as they should be! Lord Louis leaves bridge. CHEQUERS now, after her speedy start to mooring, seems to be taking an inordinately long time to secure."

CHEQUERS was rostered for a two week Care & Maintenance period in Malta dockyard in 1953, Ian remembers:- "CHEQUERS was approaching her berthing position too fast, and seemed likely to hit the graving dock caisson ahead. Dangerous this since the dock was dry and a ship was on the blocks having plates repaired. Order given 'Half Astern both', way was coming off but a ramming still looked possible, now, urgently, 'Full Astern both', the screws bit with a whoosh, the bridge staff leaned forward to a man, the foremast shook, in fact the whole ship shook, soot flew out of the galley funnel, and oily water cascaded over the jetty party. Things improving when 'Full ahead' given, at this everyone leaned backwards - not from choice - more soot from somewhere, ship shook and shivered again. Two catamarans bobbed about, rising almost up to the dockyard wall level and back again. Berthing party not well pleased. Faces of bridge occupants now a picture to behold. Apart from the Pilot's notebook being required, nothing further seemed to happen as a result of this incident.

CHEQUERS had some involvement in the Egyptian crises in the time of Mohammed Neguib, generally in the Great Bitter Lakes and Port Tewfik areas. In November 1954 she went into operational reserve at Portsmouth, having served 9 years mostly in foreign parts.

The breakers oxy-acetylene torches finally got the better of her in late 1966 at Newport.

———————◆———————

Music lover: Hears a prima donna singing in her bath, puts his ear to the keyhole.

CHIVALROUS

CHIVALROUS was a CH Class ship which had a very short life with the Royal Navy. She moved down the slipway at the yard of William Denny, Dumbarton in June 1945, and was completed for service in 1946. During this early post-war period the need for ships reduced drastically, however, destroyers and other small warships still had many duties to perform.

Joining the 14th. Destroyer Flotilla based at Malta, she was soon on immigration duties in the Levant area. Derrick Bowen was a member of the ship's Electrical Department, and he recalls that they usually operated in company with CHIEFTAIN whilst going about their immigration control business. He recalls split heads, bruised stomach and genital areas, and one gunshot wound to the forearm, all sustained by members of the boarding party carried by CHIVALROUS and formed from her own ship's company.

Moored stern-to at Haifa he recalls that the Irgun Zvei Leumi organisation had succeeded in severely damaging British merchant ships. R.N. vessels therefore took measures to ensure that they in turn did not suffer that fate. Small explosive charges were thrown overboard at irregular intervals and at the discretion of the Officer of the Watch to discourage frogmen from attaching limpet mines to the hulls of warships. Motor cutters regularly patrolled day and night carrying out the same duties. Sleep disturbance resulted from all these activities, but better tired and alive than the alternative!

Derrick rigged a jury lighting system on the upper deck of his ship to assist in these security arrangements. Inevitably, to people who served there, the parallel with the present situation in Northern Ireland is easy to see, any British serviceman was fair game to the terrorist. When going ashore, each carried smallarms.

CHIVALROUS carried out duties at the Northern end of the Red Sea in addition to the Trieste patrol. The French and Italian coasts were visited from time to time when other duties and exercises allowed. H.M.S. MESSINA a Landing Ship Tank served in company on occasion.

In mid 1954 CHIVALROUS was sold to Pakistan, renamed TAIMUR, she was finally scrapped in 1961.

◆

Pop song. A song that fortunately is not popular very long.

CONCORD

CONCORD is a nice word, it smacks of cordiality, agreement, harmony between persons and things. At first sight a strange name for a warship which has an arsenal of destructive weapons at it's disposal. On reflection however, a British warship is not going to open fire, let alone destroy anything or anybody without good cause, and a wholesome reason for so doing!

Originally to be named CORSO, H.M.S. CONCORD was commenced in November 1943, launched by Lady Leighton wife of the then Commander in Chief Portsmouth Command on 14th. May 1945, she took to her natural element at the Woolston Yard of Messrs. Thorneycroft Ltd.

First commissioned into the Portsmouth Flotilla in December 1946 she sailed for the Far East on 18th. of April 1947. She was destined to serve Britain in foreign waters for 10 years.

Working up took place in and around the Maltese Islands, moving on via Port Said, Aden, Columbo, Trincomalee, Singapore, Johore and finally Hong Kong the main fleet base in that area.

Part of the 8th. Destroyer Flotilla, she was to visit a great many ports in her time, and the list would make guests on a one year round the world liner quite envious.

It will be recalled that in April 1949, H.M.S. AMETHYST was fired on and trapped in the Yangtze River, suffering many casualties during that event. On 31st. July, Commander Kerans her Commanding Officer made a dash downstream for freedom, a most risky enterprise, but well worth the effort. She made signals at intervals on the way out toward the sea, and several ships were alerted to give covering fire, one of these was CONCORD. At the mouth of the Yangtze lies the very large communist island of Woosung, several large gunsites existed around the coast, and the communists were very quick to fire on anyone to defend their new-found territory.

These guns did open fire on AMETHYST as she steamed at full speed downstream, suddenly the wonderful moment arrived when she and CONCORD sighted each other, there was much mutual cheering as the ship's neared each other, and one member of AMETHYST'S crew said he believed it to be the finest moment of his life.

Off the Chinese coast, but further South than Woosung the shore batteries opened fire on CONCORD steaming about her lawful business. CONCORD returned the fire. This event occurred in 1950. The ship carried out all the usual patrols and exercises, shipping was often interfered with in the strait between Formosa (Now Taiwan) and the mainland of China, and the fleet responded to calls for help on many occasions.

The Korean war was in full swing in 1952, and in common with the allied fleets, CONCORD was used for patrolling and bombardment of shore targets to assist the land troops. On 23rd. April she was struck on 'Y' gun turret by what is believed to have been a 75mm shell from the area of Songjin, two men were killed and five injured. Naval honours were accorded the dead who, out of necessity were buried at sea, (returning to land in wartime conditions was not an option). In the same year, CONCORD was rammed by a Republic of Korea ship, sustaining damage to her Port Quarter.

Anti-terrorist operations in Malaya occupied the fleet for several years and CONCORD played her part in this element of fighting Communism, this time in a British controlled country. Coastal patrols and gunfire support to the land forces were the main ingredients.

In 1953 Commander C. A. James assumed command and a new ship's company joined her in August. Typhoon Rita was negotiated in September.

Immediately after commissioning CONCORD went to the assistance of M.L. 1323 which suffered damage and casualties when communist shells struck her in the Pearl River estuary, seven dead were recovered and next day CONCORD returned to retrieve the motor launch.

Christmas was spent in Kure, Japan, in company with COSSACK and COMUS. During the year she had become 'Cock of the fleet' and came second in athletics.

Many exercises were held in 1954 and at one stage CONCORD helped screen an American aircraft carrier.

February brought a further re-commissioning ceremony, this time at Singapore. Commander A. J. McCrum became skipper. Tony Nuttall of Nottingham recalls a great deal of bombardment of Indonesian terrorists thus assisting the land forces once more. Fall of shot was observed by military aircraft, by this means much more accurate fire should normally ensue. Again much exercising went on, in particular with Australian and New Zealand units. The Formosa patrol was there of course, turning up like a bad penny.

1956 saw CONCORD in the Southern Pacific carrying out duties relating to the atomic weapon tests at Montebello. Tony remembers:- "We got ashore and found complete desolation, from time to time men in white coats (perhaps most appropriately) would appear and disappear into holes in the ground. Civilised rabbits was the thought in our minds". Whilst in the area someone hooked a tiger shark which was instantly returned to the sea since it may have been contaminated. Sailing to take up her allotted station CONCORD was hit by a hurricane which holed her, CONSORT appeared to relieve her of her duties and she made it safely to Fremantle where emergency repairs were carried out sufficient to enable her to steam to Hong Kong and dry-docking.

CONCORD'S 7th. and final commission began on Wednesday the 20th. June 1956 at Hong Kong. Commander F. J. Marryat in command. Over 30 ports would be visited and a great deal of patrolling and exercising carried out. At one stage she exercised with the cruiser NEWCASTLE a species of ship which was certainly endangered by the march of progress.

Mike Taylor of Kings Lynn was serving in her when the ship visited New Zealand, some 30 of her sailors paraded through the town of Nelson situated in the Bay of Tasman on the North shore of South Island. The Mayor took the salute, and CONCORD'S C.O. was his guest on the saluting base. Local press reports spoke of destroyers being a novelty in those parts, whilst others said that they had never before seen a ward-room painted in lollipop pink. Rumour had it that the bulkhead walls of the Navigators cabin were of the same delicate shade of pink, but the deckhead was a deep blue, with, - wait for it:- stars painted on it...... Well, well, well......

The skipper had served in the destroyer KIMBERLEY during the evacuation of Crete in world war two. His ship had taken many New Zealanders to safety, and, being invited to a reunion of the New Zealand 23rd. Battalion, he was able to meet some of the men he helped rescue.

In November 1957, CONCORD berthed at Portsmouth to go into extended reserve. Against her gangway a notice appeared:- This desirable habitat for sale. For particulars ring WHItehall 9000 between 12 o'clock and Noon. That number was of course the Admiralty.

CONCORD performed further service in home waters when she was based at H.M.S. CALEDONIA the stone frigate at Rosyth. Used for training and latterly as an accommodation vessel.

On 22nd. of October 1962 her death knell was sounded when she was moved to Inverness for breaking.

◆

CONCORD. 'Y' gun mounting struck by communist shell at Songjin, Korea. Unfortunately 2 men were killed and 5 wounded by this event.

Peace. In international affairs, a period of cheating between two periods of fighting.
Ambrose Bierce.

COCKADE

In a two week period in March 1943, 21 escorted ships in convoy were sunk. These large losses emphasised the need for more escorts to give greater protection to the merchantmen whose losses were grievous.

One of the responses to this was to lay down the keel of H.M.S. COCKADE on 11th. March 1943. Viscountess Weir carried out the launching on 7th. March 1944, and the ship was fitted out and ready for service by 29th. September 1945, thus just missing the final throes of the Second Great War.

COCKADE was the first ship in the Royal Navy to be so named, and she was 'adopted' by the town of Brighton to replace the previously adopted ship, another destroyer - KIPLING - a ship with a wonderful war record, sunk by enemy action. The practice of adoption was very widespread during the war.

As the final touches were made to the ship and the crew were making ready for her first commissioning, British troops were still involved in fighting in certain limited areas where civil unrest caused difficulties to friendly governments. Java was one such land since the Dutch, occupied for four years by the Germans and therefore very weak, had no resources to enforce law and order. It fell to the British to provide that support, and this they did, actually using some Japanese prisoners to police outlying regions.

Not surprisingly, COCKADE sailed for the East, where, on 1st. February 1949 she sailed from Tokyo to assist a merchantman in distress some 300 miles away. Visiting many countries including Australia and New Zealand in mid 1947 in addition to exercising alone and in the company of other South East Asia Treaty Organisation vessels, she was ordered home and sailed for the United Kingdom, leaving Hong Kong on the 18th. November 1947. Quiet months followed her homecoming until her sister ship COMET was damaged on her passage out to the Far East when she collided with the CHEVRON whilst on night encounter exercises in the Mediterranean. COMET returned to Britain, her crew took over COCKADE and in July 1949 took her to sea and carried out trials until damage to the Starboard High Pressure Turbine forced her back into dockyard hands. On 26th. November she put to sea for compass swinging and D.G. ranging trials followed by engine power and Foxer trials two days later.

December arrived, and on the first day of that month COCKADE sailed for her new station. Unsurprisingly working up took place with Malta as the base harbour which was the base for CHEQUERS her half-sister on which served Lieutenant Prince Philip. Staying at Pieta was Princess Elizabeth who took the opportunity of visiting COCKADE in Sliema Creek.

Suez, Massawa, Aden, Trincomalee, were called at and finally COCKADE arrived at Singapore. tying up alongside another half-sister CHARITY. Course was set for Hong Kong, and, in company with CHARITY and the Light Fleet Carrier TRIUMPH exercises of a varied nature were carried out, these were marred to some extent when a Seafire, the marine version of the Spitfire, crashed on the flight deck of TRIUMPH. Fortunately for the flight deck rating who was catapulted overboard by this event, COCKADE was able to retrieve him from the water. Further exercises were held later in the year in which

TRIUMPH, COSSACK, JAMAICA, KENYA, CONSTANCE, COMUS and ships of the U.S. Navy took part. Subic Bay was one of the areas visited during this year.

Late June 1950 and COCKADE was detailed to investigate a report of a merchantman the S.S. MAXWELL BRANDER being shelled by an unknown warship. Nothing was substantiated by the destroyer, and no action could be taken.

Communist North Korean forces invaded the South of that country in 1950. COCKADE for the first time in her life was at war. Britain immediately sent other forces to join with South Korea, the U.S. and some Commonwealth forces to resist the invasion. Having an active well trained fleet in being in those waters was a wise, most useful asset to Britain and the United Nations. Ironically the first British casualties were amongst soldiers who were being carried aboard the Cruiser JAMAICA for leave in Japan. The ship had to take up her wartime role immediately she received the first signal warning of hostilities. Some sailors had also become casualties, from the shore based fire.

COCKADE carried out bombardment of railway installations, gun batteries and other targets when she arrived in Korean waters, these included the Island of Choda, and Mokpo, during August a junk was shelled and sunk, the Dutch destroyer EVERTSEN which had gone aground was towed off and passed to a United States ship who took the Dutchman to safety and a suitable port for repair work to be carried out.

Towards the end of August COCKADE acted as guardship to the TRIUMPH whilst flying off took place. 77 rounds of 4.5 ammunition was used during September, the ship herself received some splinter damage, and mines were sighted from time to time, and dealt with. Several airmen were saved from the sea, and in October more mines were located and destroyed.

October saw a major development occur when the Chinese took up the cudgel on behalf of the North. United Nations forces now faced the original enemy and the might of Communist China, which included air and sea forces, plus a massive military strength. Stringent anti aircraft and anti ship measures therefore were the order of the day amongst the allied services. COCKADE and the other Royal Navy ships had a peacetime complement, wartime conditions demanded more concentrated working conditions and the officers and men had a very strenuous time particularly in fog or storm conditions when sometimes simply living - as opposed to fighting - a ship can be a trial. It would take time before reservists and others could be made available to each ship.

Moored and floating mines were laid by the enemy, the floaters would go wherever wind and current would take them. Both had to be respected and dealt with when possible, and the old fashioned use of rifle, bren, and oerlikon was re-introduced. Clearly minesweepers had a lot of work to do in those waters, and the fleet was not let down in that respect.

Rescuing of allied airmen was a continuing role for the warships, one British pilot ditched close enough to COCKADE for him to be seen indicating he was in a position to be rescued, suddenly his aircraft disappeared below the sea and he was lost, a sad and melancholy experience for COCKADE'S crew.

It had long been recognised that the lighter short range weapons systems were not adequate to deal with determined aircraft attacks - kamikazes in the last war had demonstrated that, they had to be knocked out of the air otherwise they became 4 ton + bombs. COCKADE had her 20mm oerlikons, and 2-pounder pom-poms removed, and was fitted with 4, 40mm

Bofors guns.

The British fleet in the Far East had been able to offer immediate and most valuable help to South Korea and the Allied forces when the North attacked, and this assistance was carried on up to the cease fire and beyond.

Working up exercises took place during August 1956 when the ship re-commissioned in Singapore, on completion COCKADE joined with her sister CONSORT and the cruiser NEWCASTLE in escorting the Royal Yacht carrying Prince Philip on his Pacific cruise. Hong Kong, Singapore, Australia and New Zealand were visited, in addition to the Gilbert & Ellis Islands, Fiji, Manus and Nauru.

Duties on the Malayan West coast were carried out in 1957, visits to Japan and Korea, exercises ad infinitum, Ceylon, now Sri Lanka suffered flooding, and COCKADE rendered assistance to the civil authorities, where she could do so. On December the 19th. 1957 COCKADE finally sailed away from Hong Kong, paying off pennant flying, stem toward the South, and, after several ports of call sighted England and paid off into reserve. Breaking up took place in 1964.

◆

COCKADE in Grand Harbour, Malta. Ricasoli W/T Station in background.

"I carest not for thee Jonathan, I am suitably provided for!" Or put another way ....

56

CHARITY

J. Thornycroft of Southampton launched H.M.S. CHARITY on 30th. November 1944, just 3 days after the largest explosion ever to occur in Britain, at Hanbury, a small village 4 miles from Burton on Trent, killing 70 people and hundreds of cattle. Completed in November 1945, she joined the 14th. Destroyer Flotilla one month later.

Entering Malta, Jack Ward of Manchester who had never been abroad before, was struck by the massive ramparts that seemed to rear up all around, and despite the fact that it was winter, they exuded a warm feeling from their honey-brown surface. He says "Warships were moored everywhere, I had no idea how large the harbour was, but we seemed to need a shoehorn to find our way through" "Bosuns pipes were working overtime, attention on the upper deck, face to starboard, attention on the upper deck, face to port, carry-on, other ships were returning these salutes but eventually we found our berth. Finished with main engines".

An old hand, (3 badges on his left arm told me so), but he really *looked* old, and yet I now realise he was probably no more than 32 or 3, pointed in the general direction of the town and said to no one in particular: 'ope's we get'n shore tonight, aven't 'ad a bottle of blue in 3 years. Couldn't find any ambit last time either, Jerry saw to that'. Thinking hard to decipher his remarks I decided not to ask him to explain". When Jack eventually looked across at the other side of the harbour it reminded him of home, bomb-sites everywhere, but many men working away to rebuild and refurbish.

Jack's attention then turned to the brightly painted boats that were approaching, each with a long painted stem and stern post, and an eye on the bows. One man was standing facing the bow and pushing the oars, how strange he thought, but they looked very competent!

Working-up were the 'In' words, Jack didn't know quite what it meant but got the general drift of a lot of hard work to come, and the 'Andrew' getting it's pound o' flesh. Most people had helped paint the ship in a lighter grey, and he thought that may be it but events soon proved him to be mistaken. Much cleaning and polishing went on, he still was not sure who the man with crossed anchors on the left arm of his number 8's was, and who seemed to have a 78 for a voicebox which regularly repeated "All brightwork must be bright". He seemed a mean man, not too tall but very strong and broad, with a face like teak. Jack allowed himself a flicker of a smile when he saw him, it reminded him of a comic inscription they had on the mess notice board.

Traditionally, when leave is announced, whether in print, or by daily routine notice, it will say something like:- Leave to the First Part of the Starboard Watch, from 1600 to 2230. Chief and Petty Officers 2300. Always there would be the preferential treatment that the higher rating is accorded. Charles Atlas advertisements were widespread at the time, a picture of this wonderful specimen of manhood would be shown, complete with a deep tan, and underneath the words: "I can give you a new body in 10 days". The mess wag had put under that script:- Chief and Petty Officers 9 days.......

Action Stations, Defence Stations, working in compartments without lights, or with the bulkhead mounted lights which reminded Jack of miners lamps, which came on in an emergency, and were just enough to see by. Everything was tested and re-tested, guns were fired, and torpedoes, on one occasion at the same time, after 18 days, the Captain of the Flotilla boarded, watched two events, inspected the ship and her company and went on his way. One of the many jokers on the mess remarked "And he didn't even ask me if I was flippin enjoying myself".

"Finished playing silly buggers at last" announces a killick with the air of one who knows, 20 minutes later they are all at action stations once again, it seems that Captain D.1. was not entirely satisfied with what he'd seen. "Stand down". "For exercise, Damage control Party close up, muster port side, aft".

CHARITY took her turn patrolling the Levant, in company with CHAPLET, two illegals were boarded during May 1946. 1947 and the Palestine Patrol once again, now it seemed a dirtier game altogether and the powers that be had sent men for specific boarding party training whilst in Malta, or, so his oppo informed him, in Cyprus.

Beirut, a peaceful community without the desolation and terror of the 1980's, was a time for some relaxation when men could - in the main - sleep alnight since only a handful would need to carry out night watches. Two days into her stay and Jack walked up on deck to see a man setting up a large camera mounted on a substantial wooden tripod, intrigued, Jack and the other lads watched as the camera was carefully pointed away from CHARITY and focussing commenced, the black cloth head cover was removed from a case, anti climax, the photographer began to pace up and down on the jetty cigarette in hand. Suddenly all was revealed, a figure in what resembled a military uniform, complete with high riding boots, a Sam Browne type belt and a peaked cap with magnificent decoration and badge at the front appeared from nowhere. The olive skinned figure raised an immaculate leather gloved hand to the photographer, indicated he wanted the camera some 4 feet from it's present position and watched imperiously as this was done.

"It's Mussolini - thought he was dead". This from one of the stokers. "Is he police or soldier", from another. "He's a pompous little bastard" said someone in the second rank which had now gathered to observe the performance. "'s'not Musso, 'e 'ad a littul round pillbox 'at, 'sides didn't have any fungus on his top lip". The first picture was taken, "Musso" adopted a somewhat dramatic pose for this, followed by an even more dramatic one with a slightly aggressive touch to it.

From somewhere above the iron deck level someone shouted "Bravo", at this "Musso" pushed his chest out even further, did a strutting circle, and presented himself to the camera again. By this time the assembled CHARITY lads were thoroughly enjoying it all, tittering turned to laughter, somebody clapped and received a bow in return. "Musso" was clearly enjoying it all and seemed oblivious to the extraction of the proverbial. Another shot by the photographer, and "Musso" gave the audience a salute and marched away beaming with pleasure. To this day Jack has no knowledge of who he was, and what sort of outfit he belonged too.

May 1948 saw the end of the British Mandate in Palestine. So ended a most difficult and costly operation. Britain's armed forces were stretched to the limit, we had lost a lot of men during the various campaigns, and were practically insolvent.

CHARITY was re-commissioned at Malta in 1949 and travelled Far Eastwards in August with H.M.S. TRIUMPH to join the 8th. D.F. Christmas Day found her in the Yangtze area, myriad jobs that a destroyer is required to do followed until the Korean War began.

All the training and preparedness for war that had been carried out by the ship and her company would now be put to the test, she was to carry out all the activities that fall to the lot of a destroyer at war, and included in those was a bombardment of Inchon Harbour in company with her half-sister COSSACK, and the cruisers BELFAST, and KENYA. On a historical note H.M.S. BELFAST is preserved and may be visited in the Pool of London. In four hours, it is claimed that 415 shells were fired into the area. In December 1951 the Daily Express reported that the Western Isles of Korea were effectively kept free of communist forces by the combined efforts of the Royal Navy and Royal Marines.

Hostilities ended in Korea in July 1953. Britain once again had played a major role, whether the world at large appreciated that effort is a matter for conjecture. By 1955 CHARITY had carried out a sizeable refit and went into reserve. She was sold to Pakistan in 1958, and all old CHARITY'S will hope that they took good care of her.

◆

Boarding: Sailor scrambling aboard.

Are you going away with no words of farewell? ....

CASSANDRA

Job number J.11001 - a destroyer - was ordered to be built at Yarrows shipyard, Scotstoun. When the keel was laid down on 30th. January 1943, the first daylight raid was made on the City of Berlin. On completion at the end of July 1944, TOURMALINE, now renamed CASSANDRA joined CAMBRIAN and CAPRICE in the 6th. Flotilla of Destroyers based at Rosyth and Scapa Flow.

On 1st. November 1944 she and several sister ships escorted the liners SCYTHIA and EMPRESS OF AUSTRALIA, together carrying some 11,000 Russians, released from their German captors by the allies in Europe. The convoy arrived safely at Murmansk on the 6th. November.

On the last day of the same month she left Scapa Flow to form part of the escort to a further convoy to Russia. All the 30 merchant ships arrived safely, despite the fact that U-boats were known to be concentrated in the sea off Murmansk.

28 merchant ships with their escorts sailed for the return trip to the United Kingdom on December the 10th. One U-boat, believed to be U387 was sunk early in the voyage. Harold Scott had the morning watch in CASSANDRA'S engine room (4am to 8am). Half way through the watch he heard a very loud bang which shook the whole ship, the telegraph rang to the "Stop engine" position. Whilst the steam admission valves to the turbines were being closed to comply, Harold noticed that the ship was wallowing, and without power. He was not to see what had happened until mid-day when he was at last allowed to leave his post in the stricken ship. Striking the starboard bow section the torpedo had cut the bow portion of the ship away, and with it "A" gun - the foremost 4.5 - had disappeared also.

Cooped up in another part of the ship were two men on ASDIC duties, (anti submarine detection and indication cabinet). Morris Birkett from Birkenhead had the middle watch duty (midnight to 4am). He remembers that the watch was quiet and nothing untoward was detected during the whole four hours (ASDIC pings are not only returned from submarines, they also return from wrecks, shoals of fish, layers of alternate warm and cold water and the rest).

When the watch was relieved, he and his oppo made their way to the waiting hammocks, Morris slept in the Canteen Flat, and his mate in the Forrard messdeck. Morris was never to see him again.

Morris, asleep in his hammock thought he experienced a sensation of flying through the air, he woke up in a daze laying flat on the deck some five feet below where he should have been. He goes on "After laying a while trying to figure out what was going on, I heard the sound of running water, and all around me felt wet. Suddenly I thought of my lifebelt, and in a dazed state, promptly forgot about it!" Two figures appeared beckoning him towards them, as in a dream, in a while he realised they were a Scouse stoker he knew, and Lieutenant Brookes, urging him to go toward them. "I realised later how important their action was as several men, trying to find their way about the unlighted passageways of the ship had literally walked into the mangled wreckage of the fo'c'sle, or into the sea".

"Directed to the Wardroom, in action the sickbay, the doc. Lieutenant Ivey had a look at me, wrapped me in a blanket, and injected me before I passed out again". "When I next came round I thought I was in a straight-jacket, lying on the upperdeck". Someone said I and others were to be transferred to another warship, which at that moment hove into view. Feeling queasy, Morris said "I want to be sick" from somewhere a voice said "Be sick then, don't worry about the deck", a strange gesture in the circumstances existing. The sea state precluded the transfer to other ships, and whilst the skipper had thought it would be necessary to abandon ship, he now thought that she could be saved. Morris was carried up to the Captain's Cabin and, along with a shipmate, Don Abbott, was made as comfortable as was possible. He was extremely stiff, and in pain, and any movement was difficult. Morris remembers "I was spoon fed by a steward, after a couple of days I was able to move about a little. Nobody seemed to be sure whether we had been torpedoed or mined".

Arriving back at the ship's starting point, Kola Inlet, Morris boarded a navy lorry, to a temporary R.N. Hospital at Vaenga, there to remain until February 1945.

Each man was allowed to send one telegram, which must not say where they were, (Standard procedure in wartime). In due course he was to see his at home, and printed at the top was the heading "Murmansk". Recovery occurred in due course, and when fit enough he carried out some duties, and in the afternoons could walk or toboggan as he thought fit. Shipmates Pat. Patton, Jeff Roberts and Dudley Mills visited him, a Russian patient gave him a painting which he has to this day.

When the time arrived for him to return to the U.K. he went back to CASSANDRA, where he realised the desperately cold conditions his shipmates had to endure, the odd small electric heater here and there!

H.M.S. ZAMBESI was to be his transport back home. He says "None of wished to sleep on the mess-decks," and he found himself a spot outside the radar office and he settled there for the next two weeks.

Very bad weather was experienced, but one survivor slept on the upper-deck, fearful of going down below. No doubt the weather helped to keep U-boats down, and aircraft grounded, although they suffered a few air attacks. Morris says "I saw an explosion one day which was said to be a warship of the Royal Navy. I think it was the BLUEBELL". Arriving at Greenock, Morris was directed to the ASDIC Depot at Dunoon, a stone frigate known as H.M.S. OSPREY.

My dress consisted of a Merchant Navy type of outfit, a wooden box contained my only belongings, the joining routine beloved of the navy was commenced, part way around, since hearing was a most important part of an Asdic ratings fitness, I had to visit a specialist acoustics officer. He refused to see me on sight, ordered me out, saying "get your haircut, and then you will be seen" - "I could have killed him".

Survivors leave was granted, Morris was to see his telegram from Murmansk, and the letter from the R.N. Barracks at Chatham informing his mother that he had been injured. "His mother - he says - would not open the telegram when it arrived, until she had put the kettle on". "She had been a postwoman during the First war and remembered the many telegrams containing bad news she had delivered then".

Morris was later carried across the Atlantic by the liner AQUITANIA a far cry from a destroyer, where he was to pick up an 'Algerine Class' minesweeper and bring it back to the Firth of Forth where she plied her trade - sweeping mines. He does not forget the 62 men who died in CASSANDRA.

He left the service in mid-summer 1946.

The Commanding Officer of CASSANDRA, Lieutenant Leslie R.N. was awarded the O.B.E. (Military Division), Lieutenant Hope R.N. was awarded a Mention in Despatches, Acting Lieutenant (Engineering) Maxwell, D.S.C., R.N. was awarded the M.B.E. (Military), and Temporary Chief Stoker Willis also received the M.B.E. (Military). CASSANDRA along with a few of her crew remained in Russia, eventually dry-docked and temporarily repaired, she returned to Rosyth in mid-June 1945, under escort of another destroyer.

Docking facilities were in demand and she was towed to Gibraltar for remedial work to be carried out, she returned to Britain into Category C reserve.

Some modernisation was carried out in 1960, after which she joined the 8th. Destroyer Squadron in Far Eastern waters, serving there until 1965 and only returning for a short spell to Mediterranean and home waters once. Borneo waters saw her carrying out duties during that emergency, and she made numerous visits to many widespread ports during this time. In company with her sister CAPRICE, the two ships collided whilst working in the Far East, her poor unfortunate bows suffering further damage! Originally an all rivetted ship, after her torpedoing, she had a welded bow.

Paid off at Portsmouth by 1966, she was scrapped at Ward's Yard Inverkeithing in late April 1967., thus leaving only six of the original Flotillas afloat.

CASSANDRA was awarded the battle honour "Arctic 1944".

The price of Admiralty is paid in blood - *Kipling*.

CARYSFORT

Britain's fortunes at sea were at a very low ebb in early 1942, it is no exaggeration to describe the situation as critical, and the need for more convoy escorts as desperate. In this light the order for job number J6131 was given on 16th. February 1942. Elsewhere on this day 7 tankers were torpedoed. The previous day Singapore had surrendered. Britain was deprived therefore of her last remaining Far Eastern and East Indian base. In the Caribbean, 14 merchant ships were sunk, and Japanese submarines had sent 11 ships to the bottom in the South Java Sea area. Off the American Coast alone, some 69 ships had perished.

Convoy escorts were therefore priceless assets it seemed to the British Government, and so, amongst a host of other orders for equipment, H.M.S. PIQUE took her place on the books of Cammell Laird.

Steel required for the ship was transferred to the yard of J. Samuel White at Cowes, Isle of Wight, since Cammell Laird were fully extended, the keel was to be laid on 4th. May, but was delayed in the words of the time 'due to enemy action'. The yard had in fact suffered damage from a bombing raid by German aircraft. 8 days later the ship was commenced, and was now to be known as CARYSFORT, thus conforming with the initial letters CA in common with her 7 sister ships.

Eventually launched into the Solent on 25th. July 1944 she was completed at a time when many countries jumped on the band-wagon by declaring war on the two main Axis belligerents, 20th. February 1945.

CARYSFORT became part of the 6th. Destroyer Flotilla, the last of her class to join up, the war in Europe nearing it's end, she underwent some modifications to enable her to sail to warmer climes where the Japanese were the main enemy. When she arrived in Columbo, Ceylon, in November 1945, Japan had surrendered after receiving an atomic bomb on both Hiroshima and Nagasaki. CARYSFORT, her services now not in demand, returned to Britain in May 1946, and was placed in reserve.

Modernisation came her way, and she did two Home/Mediterranean commissions in the late 1950's. Malta called, with much time spent patrolling in the Cyprus area where EOKA terrorists were actively attempting to regain Cyprus for Greece, by removing the British, and attacking the Turkish community living in the island. Anti gun-running patrols formed a large part of this work, and always of course giving support to the land-based forces. After losing bases in Egypt and Palestine, Britain was to find Cyprus strategically important, not least in October and November of 1956, when British, French and Israeli forces attacked Egypt who had nationalised the Suez Canal and was perceived as being a tremendous threat to the well being of the world's shipping industry. Cyprus lay some 280 miles due North of the Northern coast of Egypt, and Port Said, at the Northern end of the Canal. Cyprus was therefore a handy jumping off base for the operation, the nearest base otherwise was Malta, some 1100 miles distant. In all, CARYSFORT claims to have spent the equivalent of two months in Cypriot waters.

During her commission of March to November 1958, the following facts have emerged, who would challenge there veracity? 6425 tons of furnace oil consumed (approximately

3 times the weight of the ship herself). Tins of Bluebell polish - 868, practice torpedoes fired - 12, anti-submarine grenades fired - 76, Squid projectiles fired - 37, signals made and received - 33,000, refuelled 15 times whilst under way. Her Captain during this time was Commander M. M. Dunlop, D.S.C., R.N.

Barry Brett of Openshaw, Manchester served aboard during the 1958 commission, he particularly remembers, gun-runners being captured and brought aboard CARYSFORT, whilst in Cypriot waters.

Commencing in 1962, she underwent a very long period of refitting at Gibraltar, before joining the 27th. Escort Squadron. Service in Borneo Waters where onshore the Indonesians were proving troublesome came her way, and the British ships were fully occupied by that conflagration.

Late 1967 again saw her in Far Eastern waters for a one-year stint, arriving back at Devonport on 14th. November 1968.

She paid off in early 1969, was sold to J. Cashmore, and broken up in November 1970 at Newport, she was 25 years old, and she left 3 surviving sister ships, CAMBRIAN, CAPRICE and CAVALIER.

♦

CARYSFORT, turning to port at speed.

I was walking through the dockyard in a panic, when I met a matelot old and grey. Across his shoulder was his kitbag and a hammick, and this is what I heard him say:
Chorus:-
Oh, a million miles I've travelled, and a thousand sights I've seen,
and I've always said good morning to the chief!
Chorus:-
Oh I wonder yes I wonder, did the jaunty make a blunder,
when he made this draft chit out for me,
For I've been a barrack stanchion, and I've lived in Jago's Mansion,
and I've always said good morning to the chief.
(Tune:-   Oh I wonder, yes I wonder, when the roll is called up yonder ....).

CAMBRIAN

H.M.S. CAMBRIAN who started life as SPITFIRE, as job number 1140 at the yards of Scotts of Greenock, and later at John Brown of Clydebank was completed on 17th. July 1944 on the day that RAF Spitfires severely wounded Field Marshal Rommel, and in the week that an attempt was made on the life of Adolf Hitler who was not universally admired by his own countrymen.

Full trials of the ship were not carried out due to wartime conditions, a normal practice since CAVALIER was used in that respect and proved, for instance stability, which had to suffice for the rest in the emergency conditions prevailing.

Commissioned into the 6th. D.F. with her sisters she was soon on Arctic convoy protection work. The November 1944 convoy mentioned elsewhere saw her first involvement in those waters. She was in company with the other escorts in Kola on 7th. December 1944, and unimpressed with the welcome received from our Russian allies, common amongst most of the visiting ships. Some time was spent at Leith in mid-1945 for repairs and refitting, by which time she was made ready for the warmth of Eastern waters, a relief to many after the cold and ice of the Arctic.

She was at Columbo in August 1945, the war with Japan was now at an end and much re-organising of British servicemen was again in progress.

CAMBRIAN returned home and paid off at Chatham in mid 1946.

From 1963 to 1968, CAMBRIAN carried out 3 commissions, serving both at home and abroad. Escort Squadrons were by then fashionable, as a consequence, she was in company with a variety of vessels during this period of her life.

Paying off for the last time in December 1968, she was taken to Briton Ferry in 1971 and broken up. CAMBRIAN was awarded the battle honour "ARCTIC 1944".

Then there were two ..... CAPRICE & CAVALIER.

Murmansk bound convoy, North of Norway. Senior Officer of escort: Commander Broome in a destroyer. Two submarines form part of the escort, one signals Broome: "In event of attack I intend to remain on the surface" "So do I" replied Broome.

CAPRICE

Job number 11000 at Messrs. Yarrow's works, Scotstoun was to have been named SWALLOW. Laid down in November 1942, whilst the 8th. Army and the Afrika Korps were battling at el Alamein in Egypt, whilst the Allied armies were landing at the other end of Africa, resulting in crippling loses to the remaining French warships, 1 cruiser, 10 destroyers and 13 submarines being destroyed, it was decided to re-name the ship CAPRICE. Fortunately for Britain and her friends, the French now scuttled the remaining 57 warships at Toulon, before the Germans took control of them.

Taking to the water on the 16th. September 1943, a week that saw equally momentous events, including the disabling of the pocket battleship TIRPITZ at Alten Fjord, CAPRICE was made ready for service by 5th. April 1944. Full power trials produced a speed of 32.2 knots, her main and secondary armament had already been tested in March.

Working up took place at Tobermory on Scotland's West Coast. The training base was commanded by a retired Vice-Admiral, called back to the fleet and known as "Monkey" Stephenson for some obscure reason.

For 3 weeks the ship was given every conceivable exercise, and a few which it was said only Stephenson could dream up! Driven remorselessly all day, they would get little respite during the dark hours, particularly since the Admiral's favourite trick was to try to sneak aboard in the dead of night, throw a firecracker, and inform that tired body of men - the crew, that they had been mined, port side, forward. Waiting and watching for the reaction and passing judgement after the C.O. had done whatever he thought necessary. Orders, often given in quick succession would include: Aircraft bearing Green 20, Seaboats crew, pick up survivor in the sea, Gunfire control system inoperative, 'X' & 'Y' guns crews exercise. Quarter deck sub-lieutenant take command of ship. Forced draught fan out of service. An unending list of things that could, and would happen in wartime, more so than peace.

Near hatred for the man gradually turned to a grudging admiration as the newcomers in particular saw that in those three weeks they had learned more than a year at sea would have given them, and without learning "The hard way", facing the enemy. However "Monkey" had the last say "You won't last 5 minutes against the Germans". In the event, CAPRICE earned the battle honour "Arctic 1944" for her service.

Either the arc of traverse was incorrectly set, or the mounting of the Oerlikon on the Starboard side 'whipped' slightly at the end of the traverse, when upon being fired it clipped the ship's structure and caused some damage. After further drills, "fire in the engine room, aerials shot away, re-rig aerials, main steering motor out of service", they were ready to go. Scapa Flow where they would muster to escort convoys to Iceland or Russia, Liverpool and Greenock where they would do the same for Atlantic and other convoys. CAPRICE was to have brief meetings with the transatlantic liners from time to time, and when in Iceland, the odd libertyman would be seen raising his titfer on the morrow, blackeyed, a cut here and there, "What did the American soldier say to you Johnson" would perhaps be the first question at the little wooden pulpit where justice was meted out, and, let it be said, praise on occasion where a man was promoted, at a different kind of event.

Several of the men stood on the upper deck transfixed as a report said, "torpedo, there, look" it seemed afterwards like a long time before anyone moved, but move they did, not that there's many places to go on a small ship when several hundredweights of high explosive are pointing right at you. Mercy be, it went underneath the ship to go on it's way where it would eventually sink harmlessly to the seabed. No words exist to properly describe the relief felt at such a time.

CAPRICE and her sisters CAESAR, CAMBRIAN and CASSANDRA went to North Russia. It paid the convoys to steer towards Iceland there to gather and take a course as far Northwards as was practicable, steer close to the icepack until they had to steer Southward into the Russian coast. This would keep them as far away from the then hostile coast of Norway as it was possible. One big snag, the icefield moved South in the winter, this ensured that a funneling of the convoy occurred as the convoys approached the Northern Norwegian coast, where German ships and aircraft were based. In one convoy CAPRICE exchanged places in the destroyer screen with CASSANDRA, 4 hours later CASSANDRA received a torpedo which at first was thought to be fatal, she fortunately survived.

CAPRICE took the surrender of a U-boat when the second world war ended in May 1945. Japan was still fighting stubbornly in the Far East, death to the Japanese servicemen was infinitely preferable to capture, many committed suicide rather than be taken alive. After a refit at Portsmouth, CAPRICE sailed, in common with many others, to the Indies, reaching Columbo in August 1945. Since the war was effectively over, the main work consisted of finding and looking after POW's and helping put down uprisings which began in the chaotic conditions of the time, these principally in the French Indo China, and Dutch East Indies. Neither the French, or the Dutch were in a position to safeguard these territories and Britain stepped in.

An uneasy peace settled on the world. The Russians until recently our allies, began a game of silly sods - 100 up, particularly in Germany and West Berlin, in the Russian Zone now. Soviet military services were built up at a great pace, they had the largest submarine fleet ever at one stage, having been close to defeat by German U-boats in **TWO** world wars, had we already forgotten the lesson when the talk was often of reducing our own forces, very modest in comparison with what "Uncle Joe" had at his disposal?

Many ships went into reserve, including CAPRICE. Modernised in March 1959 she again pointed her bull-ring toward the East, and joined the 8th. D.F. Mr. Fowkes of Lichfield served aboard her during this commission and reckons it was the happiest time of his career.

Terrorism in Kenya, Malaya, and Cyprus had occupied a lot of the British defence capability, these now thankfully had come to an end, and remaining areas of unrest were comparatively small. Britain had endured firefight after firefight since the Second World War, additionally in Korea and Suez.

CAPRICE re-commissioned once more in Singapore in March 1962. After much too-ing and fro-ing into dockyard hands, on one occasion due to CASSANDRA trying, not too hard thankfully, to separate the stern of CAPRICE from the remainder of the ship, using her fated bow section.

Visits to the South including Australia were made carrying out exercises en route and during the return journey. Australian Navy ships, H.M Ships TIGER, CASSANDRA, LOCH KILLISPORT and the RFA WAVE MASTER took part. H.M. Submarine

AMBUSH joined in and after the signal was given "Recreation rig may be worn on the upper-deck", fielded a suitably clad cricket team on the upper-deck and attempted to play!

Refitting over several weeks at Singapore found CAPRICE hors de combat when the Indonesian rebels tried to take over the British territories in that part of the world. Reinforcements were rushed there by sea and air, her refit complete CAPRICE took her part, and in the meantime a small contingent from the ship manned communications and took part in some operations ashore, notably commanding small boats on river patrol with Royal Marines and Gurkhas. The outcome of the war was success for Britain, but it took until August 1966.

Towards the Middle East CAPRICE now steered, calling at the tiny island of Gan, useful not least because we were no longer very welcome in India or Ceylon. Onwards to Suez, through the canal, and Malta. Some 30 Officers on what was known as the long gunnery course were aboard, and for two weeks a great deal of noise was made in the interest of good gunnery. After visiting Naples, and a return journey to Malta for a short maintenance period, the ship set off once again, reaching Bermuda on the 7th. of May 1963. CAVALIER guardship at Haiti was relieved and on to Kingston. Georgetown and Bermuda once more, and after Key West, CAPRICE was home in August. September brought a refit, and in November she joined the 21st, Escort Group. Seacat was fitted, the Far East visited once again in the mid-60's, and 1969 found the ship at Gibraltar in reserve.

During the early 70's CAPRICE was used for training Engineering Officers, she was placed on the disposal list in March 1973, and went to Shipbreaking (Queenborough) Ltd., to join all her sisters in whatever afterlife awaits a good destroyer, AND THAT LEFT ONE .... CAVALIER.

◆

CAPRICE. China Sea. Refueling.

CAVALIER

When first ordered on 24th. March 1942 CAVALIER was referred to as Job number J6099 and was to be named PELLEW. Previous pages will have indicated to the reader the grave danger posed to Britain by events at sea and the need to reinforce convoy escorts in addition to the myriad other naval requirements of the time. Launched at her builder's J. S. White and Company, at Cowes, she was now H.M.S. CAVALIER, and she was made ready to join the Home Fleet in Northern waters by 22nd. November 1944, the 6th. ship of her class to be ready for service. Destined to have a remarkably long life, she joined the 6th. Destroyer Flotilla as a "Greyhound of the Fleet".

CAVALIER and the other 7 CA's were of some 1710 tons originally, armed with 8 torpedoes, of 21" diameter, 4 main armament guns of 4.5" calibre, 2 - 40mm Bofors, 4 - 20mm Oerlikons in two mountings, and 2 single 20mm Oerlikons. It will be noted that these armament arrangements were to be greatly changed during her 28 years of service. Provision was made to roll and throw anti submarine charges over the stern of the ship. Stability and inclination trials were held by the ship on behalf of all 8 CA's.

Her first action was as part of a strike on enemy shipping off the Norwegian coast, codenamed 'Selenium', she also carried out protection of minesweepers in the Northern area, and patrolled whilst aircraft laid mines.

In the 3rd. week of February 1945 she sailed with two other ships to re-inforce the escort to convoy RA 64 who, returning from North Russia were under attack from air and seaborne units. The convoy lost some cohesion due to very bad weather, but this helped keep the U-boats down. Three ships were lost from the convoy.

Post war operations were carried out from the Clyde as part of the Western Fleet, refitted at Rosyth, she sailed for the other side of the world and the Japanese war, arriving at Columbo on 10th. November, 3 months too late for the main event, but in time to support her sisters and the British forces ashore in the Dutch East Indies. At Sourabaya extremists had failed to respond to signals ordering them to behave, and lay down their arms and CAVALIER, in company with CARRON and CAESAR opened fire on selected targets. Pictures taken at the time bring back scenes never again to be seen of men wearing only shorts and sandals serving a 4.5 gun out in the open of a destroyers upper deck.

Further excitement of a somewhat different nature came her way when she joined with the cruiser GLASGOW who was Senior Officer, and sailed for Bombay where units of the Royal Indian Navy were displaying unrest, and in a mutinous state. Known as Force 64, this duty lasted for over 2 months, when CAVALIER moved to Singapore Naval Base. Sailing on the 20th. of May for home, she arrived in Portsmouth on the 16th. June 1946 and was placed in reserve.

As a postscript to her work on the Indian West Coast, Lord Louis Mountbatten ended his duty as Viceroy of India in 1947, and much bloodshed followed this event. Hostility is still commonplace amongst the residents of the Indian subcontinent.

1955 and CAVALIER refitted at Portsmouth, followed by modernisation work at Southampton, ready for service with the Far Eastern Fleet where she relieved H.M.S. COMUS. Working up was - as usual - at Malta, and she went on to the 8th. D.F. which was composed generally of the CO Class. One journey was when she sailed with the commander in Chief aboard his Despatch Vessel ALERT to Saigon.

Atomic tests were carried out by a number of countries, in 1958 CAVALIER spent a long period patrolling the limits of the test area to ensure that nothing strayed into the danger zone, and discouraging those who sought to protest by so doing.

Harold Gough of Canterbury joined CAVALIER on 12th. July 1957 at Southampton and he remembers that when the testing of the atomic weapon was carried out, those of the crew not actually on watch were told to stand with their backs to the site of the explosion, said to be 35 miles distant, and ordered to turn round and look at the fireball after the detonation. Harold reckons it was an "incredible sight".

Regarding more everyday matters he recalls that both bunks and hammocks were to be found aboard, which indicates a transitory period in the life of the ship. General messing was superseded by Cafeteria messing, and he says that "whole meals were sometimes produced from reconstituted materials". Harold served at one time aboard CHAPLET another 'C' Class ship. Many visits were made, including the Persian Gulf area and the Island of Gan, where civil unrest was rearing it's ugly head. Royal Air Force personnel were based there, and it was important to protect the facilities installed. CAVALIER left at the end of August 1959. Dick Harris joined her in 1959 and stayed until 1961, he particularly remembers exercising with the aircraft carrier VICTORIOUS which would dwarf his own ship when in close proximity. Australia was visited twice and a refit was carried out at Singapore, undocking in February 1961.

The largest concentration of military forces in the Far East since the 2nd World War mustered for Operation "Pony Express" in April 1961, over 60 ships of a variety of nations took part. June 1962 and CAVALIER was in Hong Kong, she visited Japan and Korea with CARYSFORT, further operational visits took place including Japan again, and the Philippines. She returned to the scene of the atomic test held 4 years previously, she was returning from a visit to Australia when she was ordered "with all speed" to Singapore, there to pick up the troops and equipment and transport them at speed to the Borneo area where an armed incursion was being defended by the British stationed there. Malaysian Federated Statehood was proposed and was met with considerable opposition in Borneo and elsewhere. CAVALIER carried 100 men of the Gurkhas and a similar number of the Queen's Own Highlanders across the short sea route, thus providing a rapid reinforcement to the garrison.

Several hundred prisoners were placed under the guard of men from CAVALIER until further reinforcements arrived and the prisoners were handed over to the Royal Marines. When the time came for CAVALIER to leave the station for home she sailed across the Pacific via Christmas Island, El Salvador and the Panama Canal, and on to the Bahamas where Fidel Castro was exercising some influence and promising to disrupt the life of that island. Drug running was also to provide work for warships in that area from time to time. On 26th. May 1963 she finally reached her home port, Portsmouth, once again to be reduced to reserve.

Gibraltar was arranged for her to refit, on 21st. May 1964, whilst under tow she was struck by a tanker, the BURGAN and as a result lost some 25 feet of her bows. Returning to England, repairs were carried out at both Portsmouth and Devonport, before she was pronounced seaworthy enough to sail once again for Gibraltar. The major refitting and modernisation work began at Gibraltar in August 1964. During this work she had 'X' gun removed, and was fitted with Seacat missiles, she was again ready for service by January 1966. Now something of a hybrid since she had some 25 feet of welded bow section, with the remainder of the ship rivetted.

Commissioned at Gibraltar in September 1966, she rejoined the Home Fleet in November of that year. May 1967 saw her sail for the Far East, trouble between Arab and Jew ensured that she took the long route around Southern Africa to gain access into Eastern waters. By July and again in August and September she carried out patrols in the Beira area, and the general area of the Mozambique Channel and African Coast, work deemed necessary to try to prevent supplies getting to Rhodesia, a British Territory, under Prime Minister Ian Smith, (who flew fighter aircraft against the Germans in World War 2). With a population in excess of 3 million, Rhodesia was rich in minerals, and had a climate generally suited to European settlers, and agriculture. Harold Wilson the then Prime Minister of Britain met with Smith aboard the cruiser TIGER in Gibraltar to negotiate and settle, but there seemed to be no meeting of minds, and it was Royal Navy ships who were trying to bring to it's knees a part of the British Empire! There were even mutterings about going to war, indeed one prominent Member of Parliament strongly advocated bombing Rhodesia. Going to war against one of our own would have split the country, and particularly the armed services who as usual would have had the dirty job to do, right down the middle, however independence was declared, and eventually the country settled down.

CAVALIER formed part of the Far Eastern Fleet in late 1967. She exercised with EAGLE who carried out flying operations off Gan, moved to Singapore, and both ships returned to Gan for a repeat of the previous exercise. She responded to an emergency call from the Greek THEBEAN and carried the Chief Engineer of that ship to Gan and hospitalisation. Australia was visited, the Beira patrol duty was carried out once more in the company of TROUBRIDGE, this task ended on the 30th. April 1968. By the end of the next month she was in home waters, as part of the Western fleet.

Many more exercises in Northern waters were to follow including the NATO "Silver Tower" which again tested the movement and effectiveness of multi-national warships. By October warmer climes were visited once more. Southern France was the venue, not the "Hands, knees, and the other one" (Cannes, Nice, etc.) beloved of prewar matelots, but Marseilles, Toulon and a move to Italy and Naples, Naples not being everyones 'cup of tea', the old saying "See Naples and die" should be stood on it's head, I reckon, die first and then go to Naples - if you feel you must! To return to the serious business of warships, CAVALIER took part in another large exercise where ships of many nations worked together to perfect the art of naval warfare against an enemy clearly identified at that time as Soviet Russia and her satellites. By December CAVALIER was again in Devonport.

Duty took her to Lisbon followed by Gibraltar once again, where she underwent another long refit, returning home some 12 months later. The now commonplace arrangement of Home/Mediterranean commissions determined that CAVALIER should commence a further such task which she began in March 1970. David Thompson of Weston super Mare served aboard during this, her final tour of active service. David recalls that he lived in the after part of the ship, in bad weather they used an emergency galley to provide food and drink since the carrying of food trays from the main galley in the forward part of the ship, to the after end was not an option. During this commission the rum ration was discontinued, the expenditure saved being built into the seaman's pay - it is claimed! David Thompson has now left the service, he retains a keen interest in ships, particularly CAVALIER and has spent many hours assisting in it's role as a reminder to those to come, of what it was like to live in a 'C' Class destroyer. He donated his uniform and several photographs to the Trust, whilst the ship was at Brighton, he has a son who follows his footsteps as a sailor in the R.N.

CAVALIER carried out a rescue when the SAINT BRANDON caught fire off the West coast of Britain, in bad weather CAVALIER managed to tow the merchant vessel into Milford Haven.

Duties took her to the Mediterranean, in addition to home and Atlantic commitments, Icelandic Fishery Protection work, rather reminiscent of the Palestine Patrol without the boardings, and warmer weather, took up some of her time. Close ship/ship manoeuvring often provided the crew with an excitement not welcomed in all quarters.

1971 and the destroyer RAPID now converted into a fast anti-submarine frigate, Commander Snell in command, raced CAVALIER over a very long course but lost narrowly. Their ages were 27 and 28 years respectively, both venerable old ladies in warship terms. Recorded speed was 31.8 knots, nearly 37 miles per hour, the plates would be vibrating at that speed in a small ship! Placed on the disposal list (a euphemism for "Who'll buy") she continued in service and visited the Pool of London, tying up alongside BELFAST, another veteran of World War 2, although some 6 times larger than CAVALIER. Both ships with a proud and honourable record of service to the Crown and to the people of this country.

Sold eventually to a Trust set up to preserve her, CAVALIER was at Southampton, Brighton, and now Hebburn on Tyne, one of the great ship-building areas of the United Kingdom, her present owners: South Tyneside Metropolitan Borough Council. The Council hope to open the ship up to the paying public in 1994.

In outline, she looks very different from her 1944 condition, the low, sleek "Greyhound of the Sea" has been changed to a bulky, rather box-like superstructure, particularly at the after end, no doubt the old destroyer men find her appearance somewhat off-putting, however the "March of progress" must be marched. Now 50 years old, she serves as a reminder both as to her use, and to the reason she came into being - to ensure our very existence. We who are old enough to remember those days may not need reminders, we have a duty to forthcoming generations:- Live in peace, prepare for the opposite.

CAVALIER has the battle honour of "Arctic 1944".

◆

CRESCENT (CR) CLASS

14th. W.E.D.B.P.

CREOLE. Launched on 22nd. November 1945, at White's yard, she joined the 4th. Flotilla in 1946, and became part of the 3rd. Training Squadron, spending some time at Londonderry. For this work 'B' gun was removed to make way for other equipment. Sold to Pakistan in 1956 she underwent a long refit prior to transfer. 'B' gun was reinstated, 'X' gun removed, and two Squid systems fitted. Pakistan renamed her ALAMGIR.

CRESCENT. Name of ship of the Flotilla ordered, she entered the water on 20th. July 1944. Built by John Brown of Clydebank, she never entered service, but was sold to Canada in 1946. In 1949 she carried out exercises in the Far East with British and other units, in April 1949 she was in company with the Royal Navy's AMETHYST, COMUS, CONSTANCE and COSSACK. CONSTANCE met a team from CRESCENT for football at Hong Kong on 20th. April 1949, history does not record the result. The Canadians converted her to a fast anti-submarine frigate in the 1950's fitting her with an A/S mortar. Sonar equipment of an experimental nature was fitted at the expense of 'Y' gun at a later date, and she was scrapped during the 1960's.

CRISPIN. Built by White's at Cowes, CRISPIN was launched on 23rd. June 1945 as H.M.S. CRACCHER. Completed on 10th. July 1946, she went to the 4th. Escort Flotilla joining her sister CREOLE later in the 3rd. Training Squadron, this work necessitated removing 'B' gun to accomodate radio and other equipment. She represented Plymouth Command at the Coronation Spithead Review of the Fleet, on 15th. June 1953. Reduced to reserve in 1954, she was sold to Pakistan in 1956, that country renamed her when they took delivery as JAHANGIR. Squids had been fitted for this transfer.

CROMWELL. This vessel did not enter R.N. service when launched on 6th. August 1945. Originally named CRETAN, she was built by Scotts, and entered service with the Royal Norwegian Navy as BERGEN, in 1946.

CROWN. Launched on 19th. December 1945, she went to the Royal Norwegian Navy on completion in July 1946. Known henceforth as R.N.N. OSLO.

CROZIERS. Another vessel which was transferred to the Norwegians on completion in October 1946. Launched on 19th. September 1944 she became R.N.N. TRONDHEIM.

CRUSADER. This ship went to the Royal Canadian Navy in 1945. John Brown built her and the launch date was the 5th. March 1944. One gun was removed by the Canadians to enable A/S equipment to be fitted. Scrapped in 1960.

CRYSTAL. Launched by Yarrow on 12th. February 1945, she went to Norway as R.N.N. STAVANGER.

The CRESCENT group of 8 ships, built under the 14th. Emergency War Destroyer Building Programme, arrived too late to see any active war service in World War 2. Broadly, built and fitted out as CO's, only two went into R.N. service CREOLE and CRISPIN. Attempts to gain further information from the 3 governments who received these vessels from Great Britain have so far unfortunately proved unproductive.

♦

*Destroyers operating in the Mediterranean Sea.
Not a suitable photograph for a holiday brochure!*

"C CLASS" - Technical Details

Displacement is the weight of water displaced by the vessel. If two vessels each with the same specifications are built by the same contractor in adjacent slipways there will always be some difference in weight however slight! Similarly during fitting out even more discrepancies may arise. When stores, ammunition, and oil is loaded aboard then out of 8 ships, it is likely that they will each have a different overall weight.

In the wartime situation discussed here the Admiralty would not delay completion of a ship purely for the correct gun or guns to be found for the close range systems, it follows that differing close range systems were sometimes found. Main armament guns of 4.5 calibre were fitted, in the CA's they were hand operated, but the CH's and beyond had power operated weapons. All initially had 4 of these.

CA's

Length 363 feet. Beam 35' 9". 1710 tons + up to 800 tons.
Main armament, 4 x 4.5 Hand worked, Single mount, Open Housing.
Maximum elevation 55°.
Steam raising plant: 2 Admiralty Pattern Boilers, 3 Drum, Main steam outlet pressure 300 p.s.i.g. and 630°F.
Propulsion: Parsons 2 shaft geared turbines, developing 40,000 shaft horsepower: 350 RPM.
Ahead steering revolutions: 315. Hard over to Hard over, 20 seconds.
Maximum angle: (Bubble) 12½°.
2 Screws: Outward turning.
Rudder: 76 sq ft. Maximum angle 35°.
Mean draught (Foreward) 11' 4" (Aft) 14'.
Torpedo armament: 8 x 21" in two mounts of four.
Maximum oil fuel storage: 588 tons.
Maximum speed expected: 34 knots. Complement: 186 + 36 Leader.
All rivetted construction. 4 Depth-charge throwers, 2 runways.
Fitted as Leader: CAVENDISH.

CH's

Some rivetted, some welded construction. (See note over page), otherwise details as for CA's except:- Only 4 torpedo tubes fitted in order to make the provision of a power operated main armament system valid.
Fitted as Leader: CHEQUERS.

CO's

All welded construction. Basic design as for previous sub-classes. (N.B. One official document states the H.M.S. CRISPIN was the first all-welded destroyer). Six were modernised in the 1950's, COMUS & CONSTANCE were disposed of and therefore not modernised.
Fitted as Leader: CONSTANCE.

Broadly, and except where stated to the contrary, the following modernisations and updating were carried out:

Developments in anti-submarine warfare brought about a change in equipment, in general the 'C' class had 'X' gun removed and a large box-like structure fitted which supplied the A/S mortars with projectiles. This first major change in silhouette detracted from the early sleek lines of the vessels. One bank of torpedo tubes and in some cases both, were removed from the CA's in order to compensate for weight added by Gunnery control gear, and other top hamper. CHAPLET, CHIEFTAIN, COMET and CONTEST were all fitted for minelaying by conversion, this gave them a distinctive 'fantail' appearance, reminiscent of American warships. CHAPLET, in addition to 'X' also lost 'Y' gun to make way for this alteration. Several other major modifications were carried out and CARRON lost 'B' gun to make way for a sizeable foreward extension to her bridge. Navigation training was carried out by CARRON subsequently. Seacat missiles became a feature of armaments carried by the class, and by 1960 CARRON had no remaining 4.5 guns.

Typical radio gear carried aboard the class would be B28 (Marconi CR 100) receivers, B29 L/F receiver, 60EQR transmitter, T.B.L. TCS and in early 50's., B40 and B41 receivers. More sophisticated equipment was used as time went on.

◆

'C' CLASS - Pennant Numbers

CAESAR *	R 07	COCKADE	R 34 (R 74)?
CAMBRIAN	R 85	COMET	R 26
CAPRICE	R 01	COMUS	R 43 - D 20
CARRON	R 30	CONCORD	R 63 - D 03
CARYSFORT	R 25	CONSORT	R 76
CASSANDRA	R 62	CONSTANCE	R 71
CAVALIER	R 73	CONTEST	R 12 (D 48)
CAVENDISH	R 15	COSSACK *	R 57 (D 15)
CHAPLET	R 52	CREOLE	R 82
CHARITY	R 29	CRESCENT *	R 16
CHEQUERS *	R 61	CRISPIN	R 68
CHEVIOT	R 90	CROMWELL	R 35
CHEVRON	R 51	CROWN	R 46
CHIEFTAIN	R 36	CROZIERS	R 27
CHILDERS	R 91	CRUSADER	R 20
CHIVALROUS	R 21	CRYSTAL	R 38

The prefix R was changed in the early 1950's, and D was used from thereon. Pennant numbers given to the CA's before they were renamed are not shown. There are some apparent discrepancies within the CO's these are shown above. Captain's (D) often carried no Pennant Number. The prefix 'R' was used after World War 2 for Aircraft Carriers.

* - "Name" ships of Class.

GLOSSARY OF TERMS
for Non R.N. readers!

Andrew	The Royal Navy
A/S	Anti-submarine.
Bare hook	Leading rate without any long service badges.
Belay	Stand fast, wait.
Bulkhead	Wall (of compartment).
Bullring	Fairlead at extreme 'nose' of ship.
Chief Yeoman	Senior visual signalman.
Club Swinger	Physical training instructor.
Cox'n	Coxswain. Senior rating of the seaman branch. Takes the wheel when in action or on other specified occasions.
Crusher	R.N. Policeman. officially: Patrolman.
Damager	NAAFI Manager.
Deckhead	Ceiling (of compartment).
D/F	Direction finder, radio.
Draft	Moved to a new appointment.
Draft Chit	Document ordering that move.
D.S.	Destroyer Squadron.
D.S.C.	Distinguished Service Cross. Awarded to Officers only.
D.S.M.	Distinguished Service Medal. Awarded to Ratings only.
D.S.O.	Distinguished Service Order. Awarded to Officers only.
Enosis	Union with Greece.
EOKA	Cypriot terrorist organisation.
Flannel	To whitewash, mislead.
Green Rub	Hard luck.
Grey Funnel Line	The Royal Navy.
Haganah	Jewish organisation (1920) set up to gain a Jewish homeland.
Heads	Toilets, washrooms.
I.I.'s (Eye Eye's)	Illegal Immigrants (into Palestine).
Irgun Zvei Leumi	Splinter from Haganah, terrorist group.
Jago's Mansion	Royal Naval Barracks - Devonport.
Jimmy	Jimmy the one, 1st. Lieutenant.
José	Any Maltese.
Lanchester	U.S. rifle. (often found on ships repaired in the U.S. after torpedoing etc.).
L.S.T.	Landing ship tank.
Maltese Lace	Worn, ragged clothing.
M.B.E.	Member of the British Empire (May be civil or military division).
Mossad	Jewish. Military wing of Haganah, now Israeli intelligence.
O.B.E.	Order of the British Empire.
Pennant	Signal Flag. Broader at mast, tapering toward free end.
Pipe Down	Order to keep quiet, traditionally at 2200 in 1940 - 50's.

P/L	Plain Language.
Quarter	The point between abeam and dead astern.
Rattle	On remand. On report.
R.F.A.	Royal Fleet Auxiliary vessel.
Rig of the Day	Dress to be worn as ordered.
R.O.K.	Republic of Korea.
Rose Cottage	Mess set aside for those with venereal disease.
Stern Gang	Further Israeli splinter group led by Isaac Stern. Terrorists.
Stripey	Long service rating, usually with 3 stripes on his left arm. Does not denote rank in R.N.
Tailor made	Cigarettes bought ready made.
T.G. or U.A.	Temperance, Grog or Under Age. (do I get a tot) ???
Three badge F.A.	See stripey.
Tickler	Service tobacco allowance, contained in a large sealed tin, ready to be rolled into cigarettes.
T.O.R.	Time of receipt (signalling).
Trot	Area of harbour/creek where numbers of ships may moor to buoys in close company.
V/S	Visual signalling.
W/T	Wireless telegraphy (morse).

◆

Juggler working the cruise liners, parrot in the audience 3 night's running, "It's up his sleeve, it's in his top pocket," Conjurer is going quietly spare. Fourth night, 5 minutes into the act, and the ship's boilers blew. Conjurer and parrot find themselves on the same raft. Parrot watches him for 3 days, head held to left, head held to right, finally parrot says: "Alright, I give up. What have you done with the ship?"

EPILOGUE

Whither the Royal Navy in the future? Women now serve at sea in major warships, with all the attendant problems of accommodation, sex and discipline. One lady has recently received £300,000 for wrongful dismissal after leaving the service due to pregnancy, a fortune, which had it been paid to a serviceman, in equivalent terms in the 1940's or 1950's, would have led him to believe he had won the pools!

The implications of the foregoing on a service dedicated to fighting a war at sea are enormous. Previously, men would be punished for being unable to carry out work due to unfitness resulting from their own actions!

Perhaps the most disturbing fact of all is that men now no longer come forward to join the Royal Navy in sufficient numbers to man the very much smaller fleet than the country had heretofore.

Equally, the cohesive efforts of Great Britain Limited in future emergencies, perhaps would not be centred on the Monarchy, since many now, unaccountably, appear to be disaffected. They possibly forget that King George Vth., and King George VIth., both served at sea, Prince Philip commanded MAGPIE, Prince Charles commanded a Motor Minesweeper, and the Duke of York won his spurs in the South Atlantic.

Sizeable immigration has occurred since 1945 when many refugees arrived from war-torn European areas, and subsequently very many more have arrived from the Commonwealth. Recent events have shown that in latter times, people have not been as readily adaptable to our way of life as were the earlier arrivals.

Returning to the Royal Family, the superb service and devotion to duty of the Queen and the Queen Mother cannot go unremarked. Sadly I sense a tug of war over whether Prince William will serve in the Royal Navy, and in view of press attention, this can do nothing but harm to the service. All is not gloom however, and the British Services still command world wide respect.

―――――◆―――――

ACKNOWLEDGEMENTS

During the long period of gestation of the work, my immediate family have been a help, my thanks go to them, and their co-operation and assistance are hereby acknowledged.

Ernie Balderson
Tony Nuttall
Peter Carlisle
Alan Oakes
Mike Taylor
Tony Fleming
Mrs. James Smith
Hugh Jones
Les Reader
George Taylor
Frank Bentley
Barry Brett
J. Ryan
M. Jones
Eddie Barrett
Michael Cox
Dennis Barker
Norman Ambler
Bill Johnston
Charles Carter
R. J. Wren
William Burden
Gordan Banham
Arthur Aston
David Thompson
W. R. Bartle
Ian Wardle
Steve Mathis
Bill Gravenor
Derek Powell
Barry Morgan
Richard Fitch

Allan Gunnis
Roy Lidgett
Peter Saunders
Sid Anning
J. R. Badger-Smith
Jack Mead
Tony Thomas
Wally Steptoe
Cyril Allwood
Frank Woods
Charles Wardle
Alan Quartermaine
Harold Gough
Geoff. Lane
Bob Bucknall
Geoff Crawshaw
Ron Morris
Bill Cairns
Roy Millward
Reg Ward
Morris Birkett
Stuart Johnson
David Bince
Dave Killelay
Derek Bowen
Dick Harris
Bill Brown
E. J. Fowkes
Gordon Juncar
Mac Brodie
David Ashby
George Toomey
Lilian Carlisle

Bob Deal
E. Bent
Matt Griffin
E. G. Robinson
Tom Fennelly
Kathleen Mead

My thanks also to the 8th. Destroyer Flotilla Association, the 6th. Destroyer Flotilla Association, H.M.S. GANGES Association, Arctic Veterans Assn. South Tyneside Metropolitan Borough Council. H.M.S. CAVALIER Trust, and H.M.S. COSSACK Association.